T0170965

The Bones

A handy where-to-find-it
pocket reference companion
to Euclid's *Elements*

The Bones

A handy where-to-find-it pocket reference companion to Euclid's *Elements*

Green Lion Press
Santa Fe, New Mexico

© copyright 2002, 2010 by Green Lion Press.

All rights reserved. No part of this publication may be reproduced, stored in a retrieval system, or transmitted, in any form or by any means, electronic, mechanical, photocopying, recording, or otherwise, without prior written permission of the publisher.

Manufactured in the United States of America.

Published by Green Lion Press
www.greenlion.com

Conceived, designed, and edited by Dana Densmore.
Set in 10-point Stone Sans.
Printed and bound by Sheridan Books, Inc., Chelsea, Michigan.

Cover illustration adapted from Raphael's *School of Athens* by W.H. Donahue and D. Densmore.

Cataloging-in-Publication Data:

Euclid
Elements/by Euclid
Translated by Thomas L. Heath

ISBN 978 1 888009-21-7

1. History of Mathematics. 2. Geometry. 3. Classics.
I. Euclid (fl. c. 300 B.C.E.) II. Heath, Thomas L. (1861–1940)
III. Title.

QA31.E8752 2002

Library of Congress Control Number 2002110364

Contents

The Green Lion's Preface

The Bones is a pocket reference edition of Euclid's *Elements* with diagrams and enunciations for every proposition, but no proofs. Green Lion Press offers it as a companion volume to our unabridged edition of the thirteen books of the *Elements*.*

Equally at home in backpack or pocket, on desk or bedside table, its possibilities are limited only by one's imagination.

The Bones allows active users of Euclid to quickly find a proposition they need or remember without having to leaf through all the pages of proofs. Sometimes one remembers what the diagram looks like and can most easily find it by scanning the diagrams. Other times one might want to look at the wording of the propositions' enunciations to find the right one.

Another important use of this book is to get an overview of Euclid's system. The succinct presentation of *Elements* allows a reader to see the ingenious and elegant structure of Euclid's work emerge.

One can find the answers to questions about what is covered in this great work without being bogged down in its size. Just what does Euclid prove and where does he prove it? What is the range and what is the depth of his treatment of particular topics?

Yet another use, one that will be the most important use for some people, will be as a reference while working

---

* *Euclid's Elements,* Green Lion Press, ISBN 1-888009-18-7 (cloth binding with dust jacket); ISBN 1-888009-19-5 (sewn softcover binding).

through other classics of mathematics. *The Bones* is valuable for anyone reading authors who cite Euclid, and especially for readers of authors who, like Newton, rely on Euclid without always citing the specific propositions.

This handy totebook has still more enjoyable possibilities. Tucked into a pack when headed for a week in the backcountry—or a wait at the airport—it can be pulled out to offer the entertainment of trying to reconstruct the proofs on one's own.

The Bones includes an annotated table of contents with a general identification of the subject matter of each book and a listing of the number of propositions in each, along with location of all sets of Definitions.

Other features and supporting material, such as an extensive index and glossary, a bibliography, and explanatory notes, are included in our full Euclid edition, together with complete proofs of all propositions. For these aids, we refer you to that volume. This companion volume has been kept short and sweet for convenience and ease of use.

We gratefully acknowledge the help in preparation of this volume of our new associate editor Howard Fisher and of our Hodson Trust Internship Program summer intern Kathleen Kelly.

May this little book save you as much time in your use of Euclid as it is going to save us in ours.

Dana Densmore
William H. Donahue
for Green Lion Press

Euclid's Elements
Book I

Definitions

1. A *point* is that which has no part.
2. A *line* is breadthless length.
3. The extremities of a line are points.
4. A *straight line* is a line which lies evenly with the points on itself.
5. A *surface* is that which has length and breadth only.
6. The extremities of a surface are lines.
7. A *plane surface* is a surface which lies evenly with the straight lines on itself.
8. A *plane angle* is the inclination to one another of two lines in a plane which meet one another and do not lie in a straight line.
9. And when the lines containing the angle are straight, the angle is called *rectilineal.*
10. When a straight line set up on a straight line makes the adjacent angles equal to one another, each of the equal angles is *right,* and the straight line standing on the other is called a *perpendicular* to that on which it stands.

11. An *obtuse angle* is an angle greater than a right angle.
12. An *acute angle* is an angle less than a right angle.
13. A *boundary* is that which is an extremity of anything.
14. A *figure* is that which is contained by any boundary or boundaries.
15. A *circle* is a plane figure contained by one line such that all the straight lines falling upon it from one point among those lying within the figure are equal to one another;
16. And the point is called the *centre* of the circle.
17. A *diameter* of the circle is any straight line drawn through the centre and terminated in both directions by the circumference of the circle, and such a straight line also bisects the circle.
18. A *semicircle* is the figure contained by the diameter and the circumference cut off by it. And the centre of the semicircle is the same as that of the circle.
19. *Rectilineal figures* are those which are contained by straight lines, *trilateral* figures being those contained by three, *quadrilateral* those contained by four, and *multilateral* those contained by more than four straight lines.
20. Of trilateral figures, an *equilateral triangle* is that which has its three sides equal, an *isosceles triangle* that which has two of its sides alone equal, and a *scalene triangle* that which has its three sides unequal.

21. Further, of trilateral figures, a *right-angled triangle* is that which has a right angle, an *obtuse-angled triangle* that which has an obtuse angle, and an *acute-angled triangle* that which has its three angles acute.

22. Of quadrilateral figures, a *square* is that which is both equilateral and right-angled; an *oblong* that which is right-angled but not equilateral; a *rhombus* that which is equilateral but not right-angled; and a *rhomboid* that which has its opposite sides and angles equal to one another but is neither equilateral nor right-angled. And let quadrilaterals other than these be called *trapezia*.

23. *Parallel* straight lines are straight lines which, being in the same plane and being produced indefinitely in both directions, do not meet one another in either direction.

Postulates

Let the following be postulated:

1. To draw a straight line from any point to any point.
2. To produce a finite straight line continuously in a straight line.
3. To describe a circle with any centre and distance.
4. That all right angles are equal to one another.
5. That, if a straight line falling on two straight lines make the interior angles on the same side less than

two right angles, the two straight lines, if produced indefinitely, meet on that side on which are the angles less than the two right angles.

Common Notions

1. Things which are equal to the same thing are also equal to one another.
2. If equals be added to equals, the wholes are equal.
3. If equals be subtracted from equals, the remainders are equal.
4. Things which coincide with one another are equal to one another.
5. The whole is greater than the part.

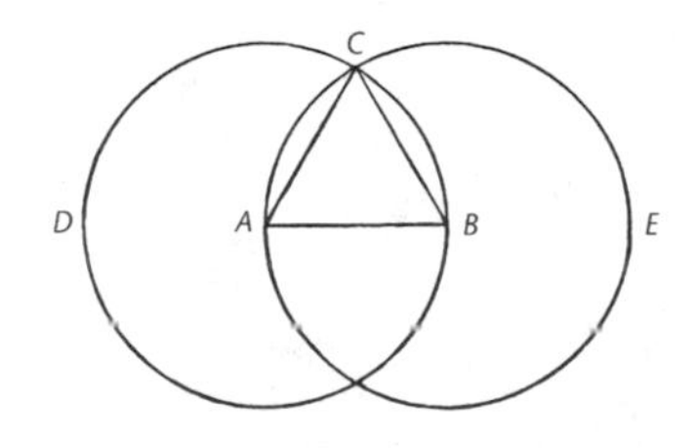

Proposition 1

On a given finite straight line to construct an equilateral triangle.

Proposition 2

To place at a given point [as an extremity] a straight line equal to a given straight line.

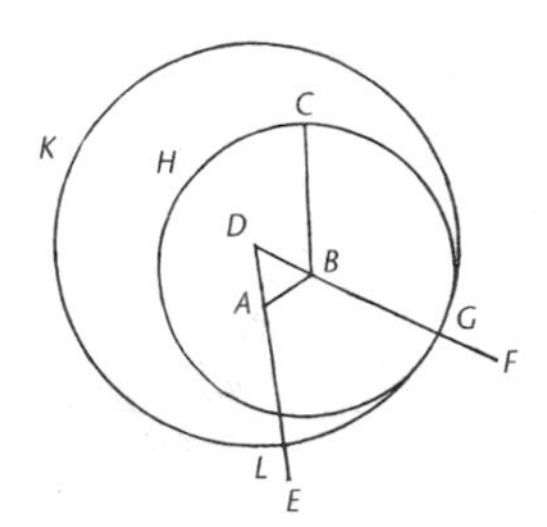

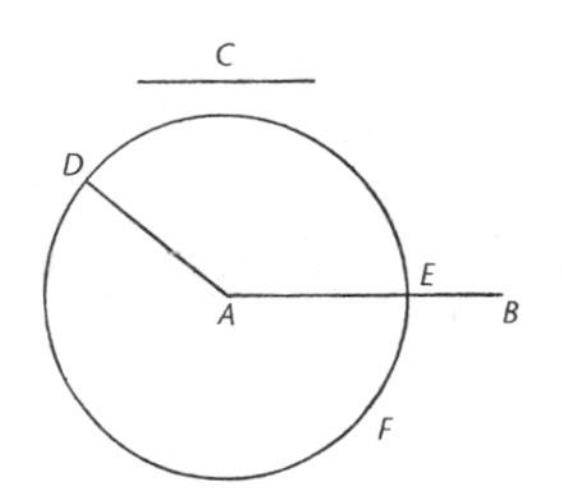

Proposition 3

Given two unequal straight lines, to cut off from the greater a straight line equal to the less.

Proposition 4

If two triangles have the two sides equal to two sides respectively, and have the angles contained by the equal straight lines equal, they will also have the base equal to the base, the triangle will be equal to the triangle, and the remaining angles will be equal to the remaining angles respectively, namely those which the equal sides subtend.

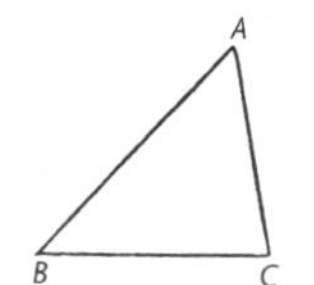

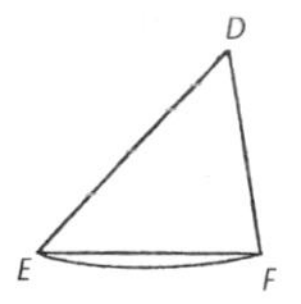

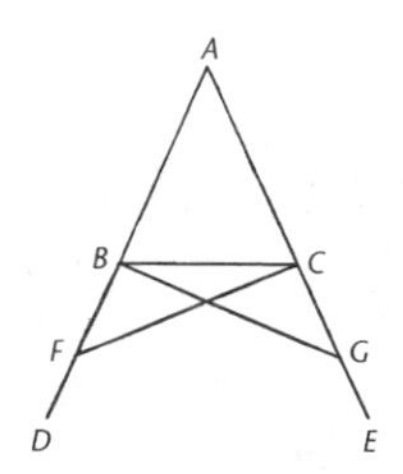

Proposition 5

In isosceles triangles the angles at the base are equal to one another, and, if the equal straight lines be produced further, the angles under the base will be equal to one another.

Proposition 6

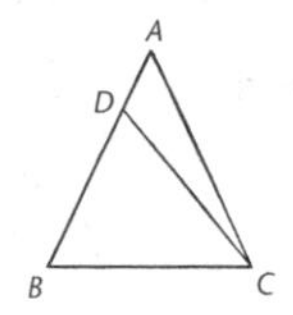

If in a triangle two angles be equal to one another, the sides which subtend the equal angles will also be equal to one another.

Proposition 7

Given two straight lines constructed on a straight line [from its extremities] and meeting in a point, there cannot be constructed on the same straight line [from its extremities], and on the same side of it, two other straight lines meeting in another point and equal to the former two respectively, namely each to that which has the same extremity with it.

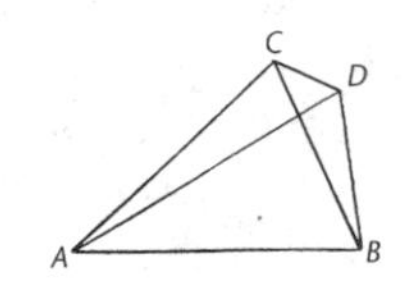

Proposition 8

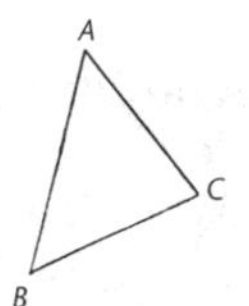

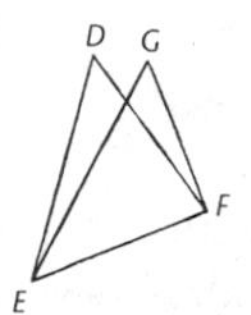

If two triangles have the two sides equal to two sides respectively, and have also the base equal to the base, they will also have the angles equal which are contained by the equal straight lines.

Proposition 9

To bisect a given rectilineal angle.

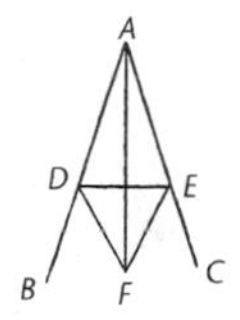

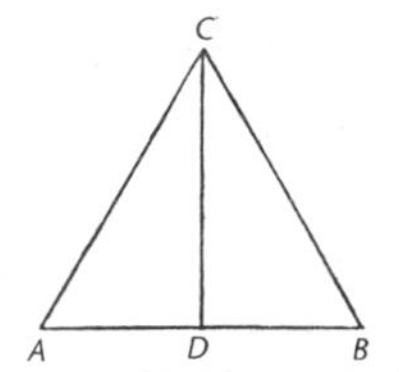

Proposition 10

To bisect a given finite straight line.

Proposition 11

To draw a straight line at right angles to a given straight line from a given point on it.

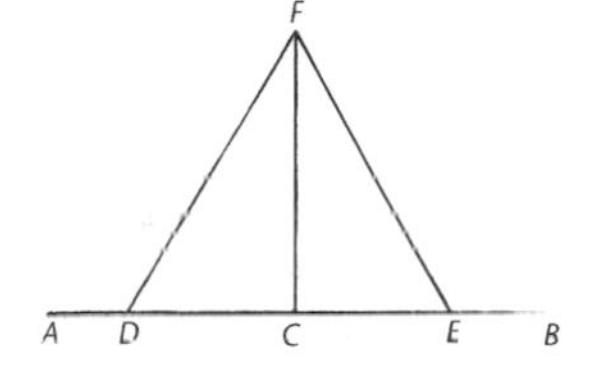

Proposition 12

To a given infinite straight line, from a given point which is not on it, to draw a perpendicular straight line.

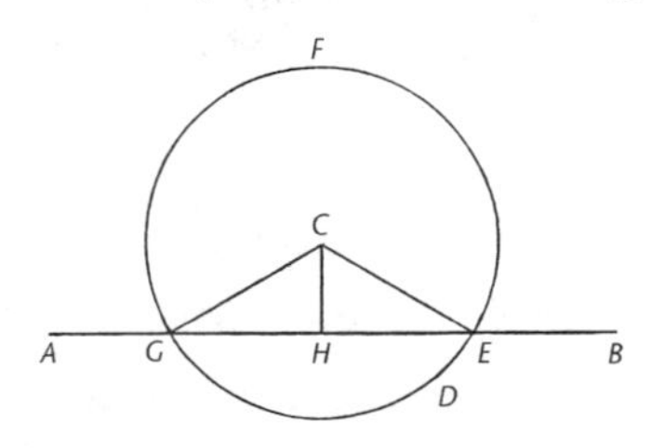

Proposition 13

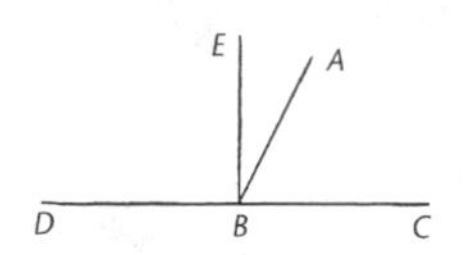

If a straight line set up on a straight line make angles, it will make either two right angles or angles equal to two right angles.

Proposition 14

If with any straight line, and at a point on it, two straight lines not lying on the same side make the adjacent angles equal to two right angles, the two straight lines will be in a straight line with one another.

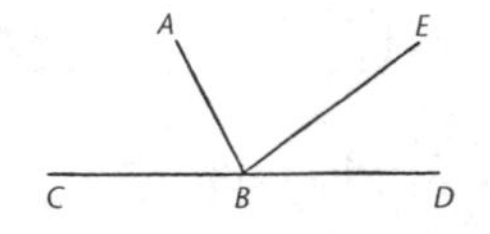

Proposition 15

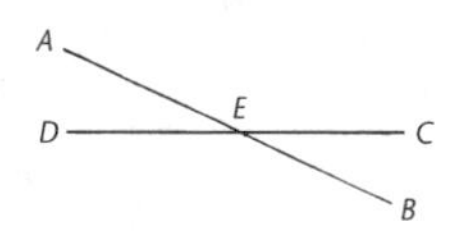

If two straight lines cut one another, they make the vertical angles equal to one another.

⟨**PORISM**. From this it is manifest that, if two straight lines cut one another, they will make the angles at the point of section equal to four right angles.⟩

Proposition 16

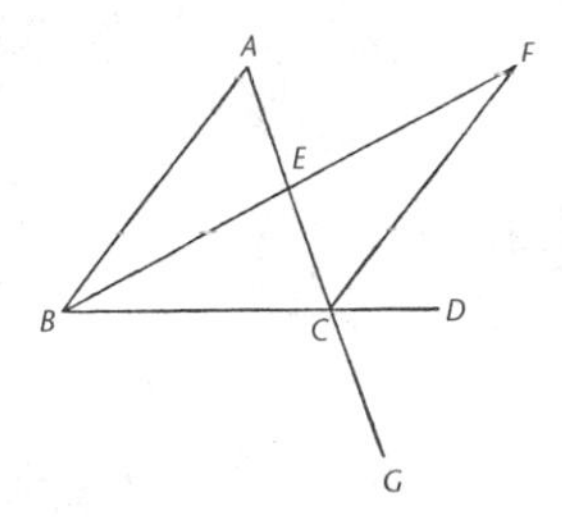

In any triangle, if one of the sides be produced, the exterior angle is greater than either of the interior and opposite angles.

Proposition 17

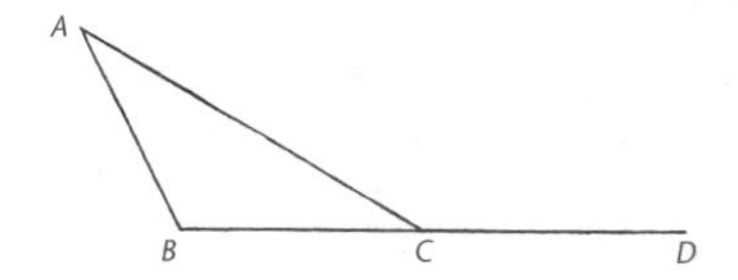

In any triangle two angles taken together in any manner are less than two right angles.

Proposition 18

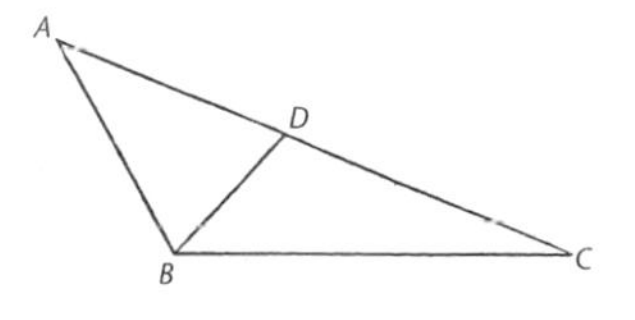

In any triangle the greater side subtends the greater angle.

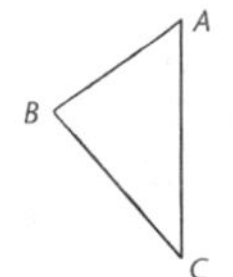

Proposition 19

In any triangle the greater angle is subtended by the greater side.

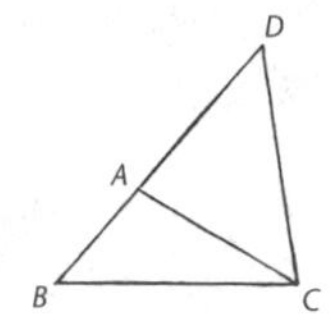

Proposition 20

In any triangle two sides taken together in any manner are greater than the remaining one.

Proposition 21

If on one of the sides of a triangle, from its extremities, there be constructed two straight lines meeting within the triangle, the straight lines so constructed will be less than the remaining two sides of the triangle, but will contain a greater angle.

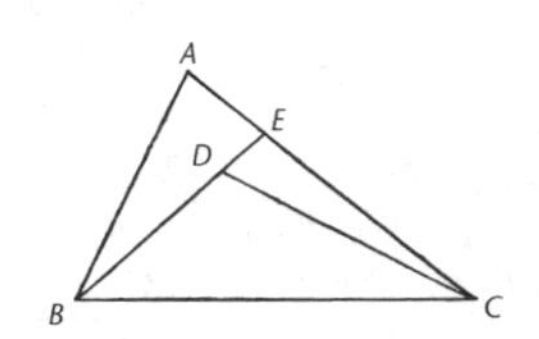

Proposition 22

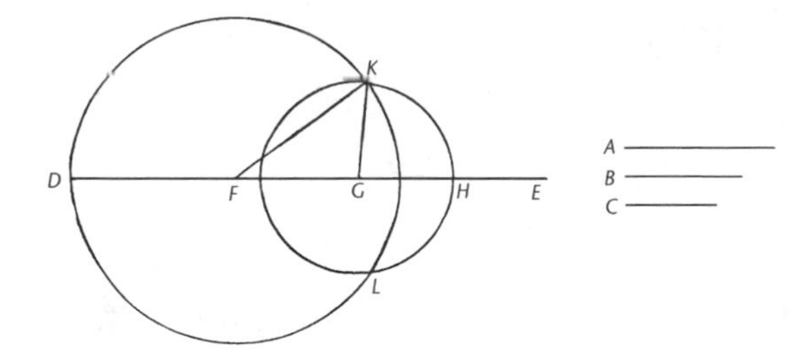

Out of three straight lines, which are equal to three given straight lines, to construct a triangle: thus it is necessary that two of the straight lines taken together in any manner should be greater than the remaining one.

Proposition 23

On a given straight line and at a point on it to construct a rectilineal angle equal to a given rectilineal angle.

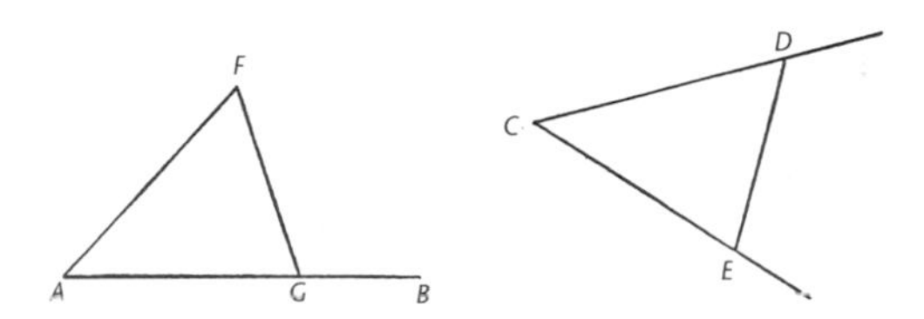

Proposition 24

If two triangles have the two sides equal to two sides respectively, but have the one of the angles contained by the equal straight lines greater than the other, they will also have the base greater than the base.

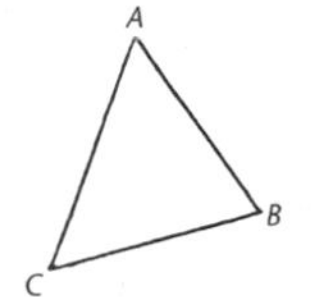

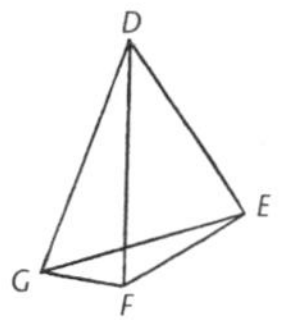

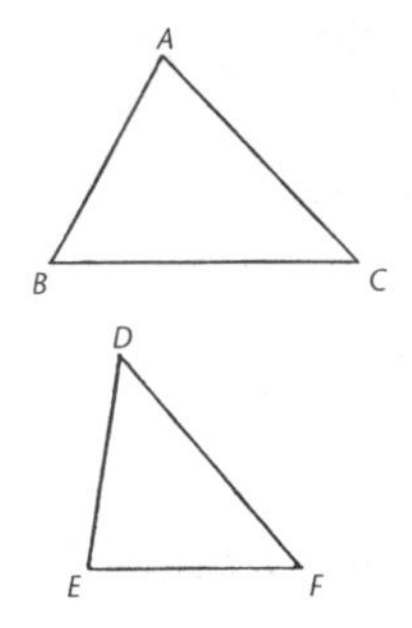

Proposition 25

If two triangles have the two sides equal to two sides respectively, but have the base greater than the base, they will also have the one of the angles contained by the equal straight lines greater than the other.

Proposition 26

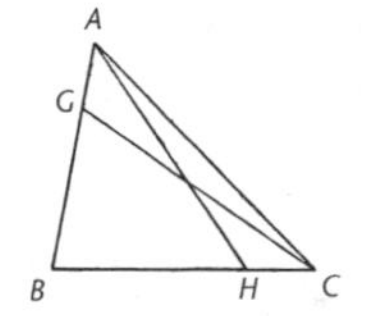

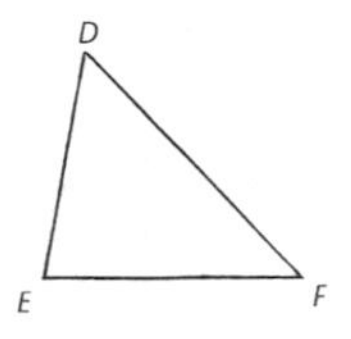

If two triangles have the two angles equal to two angles respectively, and one side equal to one side, namely, either the side adjoining the equal angles, or that subtending one of the equal angles, they will also have the remaining sides equal to the remaining sides and the remaining angle to the remaining angle.

Proposition 27

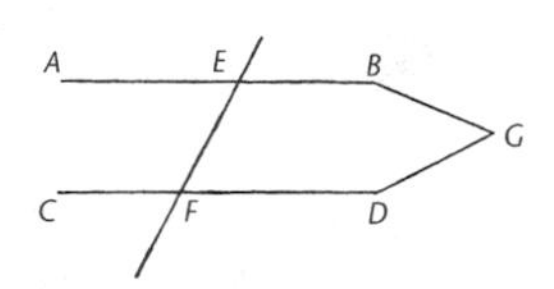

If a straight line falling on two straight lines make the alternate angles equal to one another, the straight lines will be parallel to one another.

Proposition 28

If a straight line falling on two straight lines make the exterior angle equal to the interior and opposite angle on the same side, or the interior angles on the same side equal to two right angles, the straight lines will be parallel to one another.

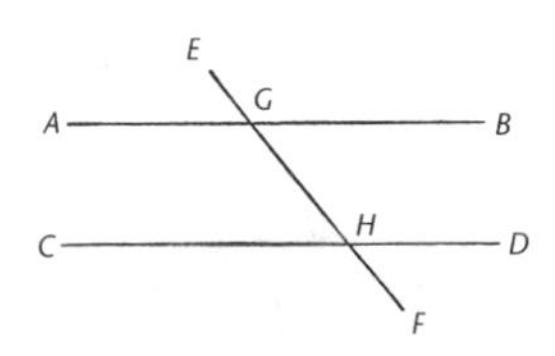

Proposition 29

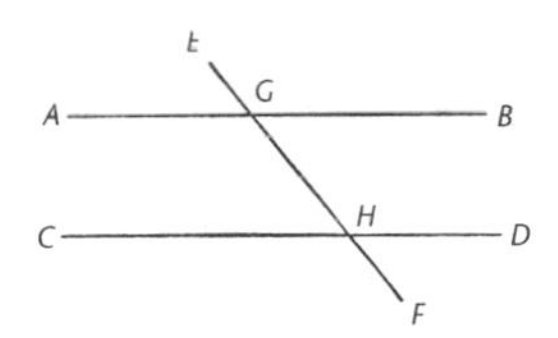

A straight line falling on parallel straight lines makes the alternate angles equal to one another, the exterior angle equal to the interior and opposite angle, and the interior angles on the same side equal to two right angles.

Proposition 30

Straight lines parallel to the same straight line are also parallel to one another.

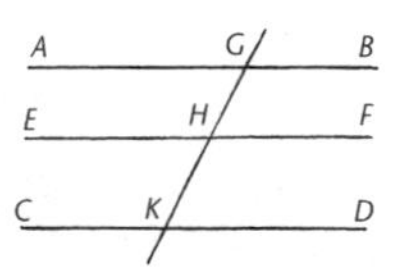

Proposition 31

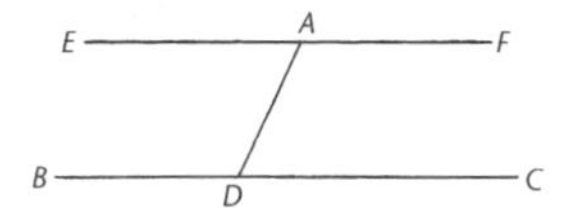

Through a given point to draw a straight line parallel to a given straight line.

Proposition 32

In any triangle, if one of the sides be produced, the exterior angle is equal to the two interior and opposite angles, and the three interior angles of the triangle are equal to two right angles.

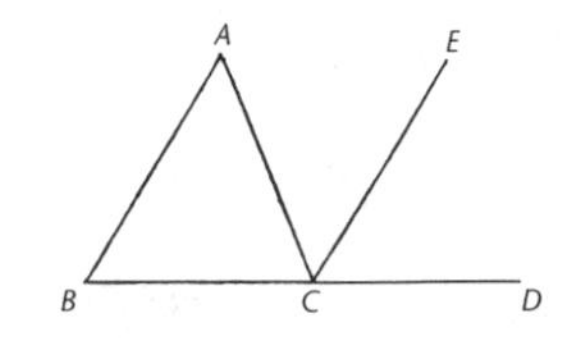

Proposition 33

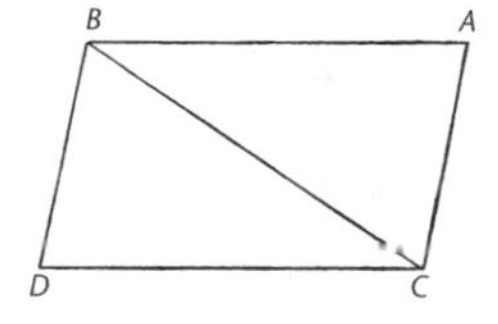

The straight lines joining equal and parallel straight lines [at the extremities which are] in the same directions [respectively] are themselves also equal and parallel.

Proposition 34

In parallelogrammic areas the opposite sides and angles are equal to one another, and the diameter bisects the areas.

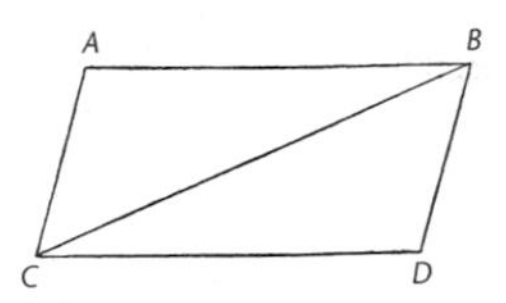

Proposition 35

Parallelograms which are on the same base and in the same parallels are equal to one another.

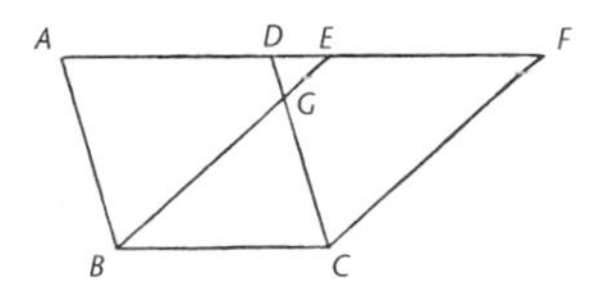

Proposition 36

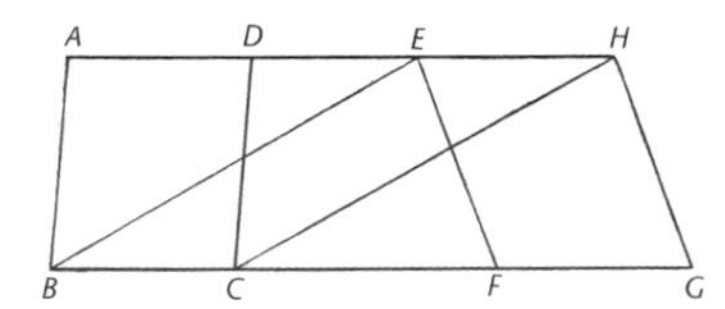

Parallelograms which are on equal bases and in the same parallels are equal to one another.

Proposition 37

Triangles which are on the same base and in the same parallels are equal to one another.

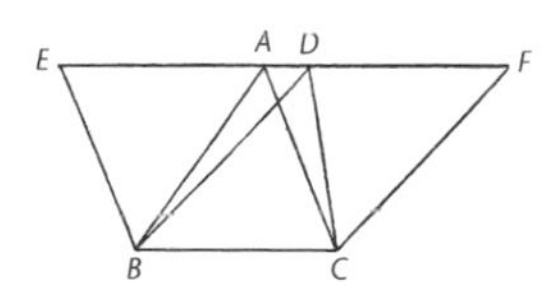

Proposition 38

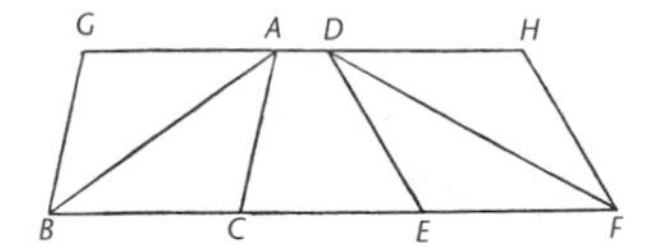

Triangles which are on equal bases and in the same parallels are equal to one another.

Proposition 39

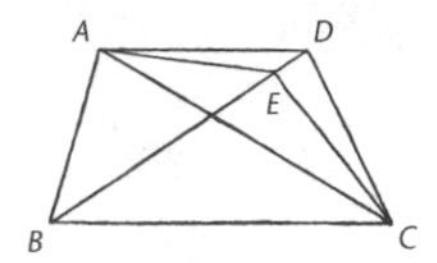

Equal triangles which are on the same base and on the same side are also in the same parallels.

⟨Proposition 40

Equal triangles which are on equal bases and on the same side are also in the same parallels.⟩

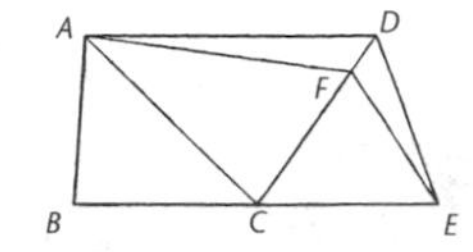

Proposition 41

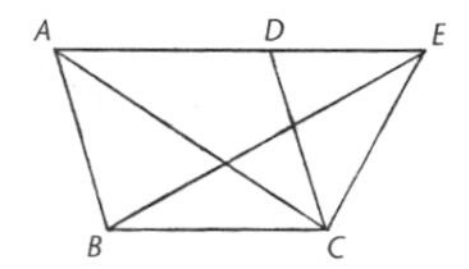

If a parallelogram have the same base with a triangle and be in the same parallels, the parallelogram is double of the triangle.

Proposition 42

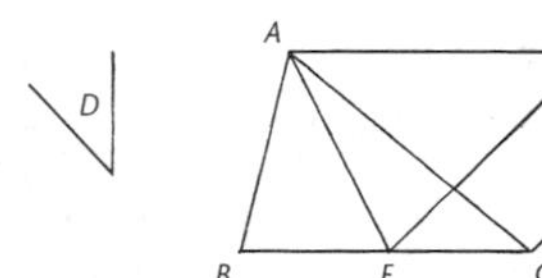

To construct, in a given rectilineal angle, a parallelogram equal to given triangle.

Proposition 43

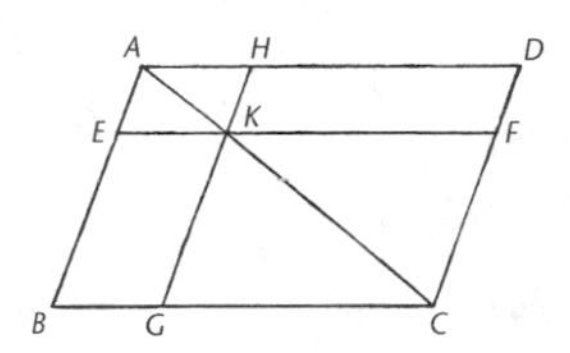

In any parallelogram the complements of the parallelograms about the diameter are equal to one another.

Proposition 44

To a given straight line to apply, in a given rectilineal angle, a parallelogram equal to a given triangle.

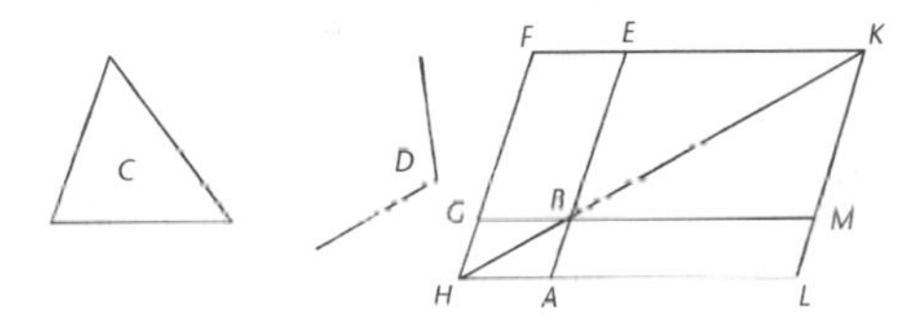

Proposition 45

To construct, in a given rectilineal angle, a parallelogram equal to a given rectilineal figure.

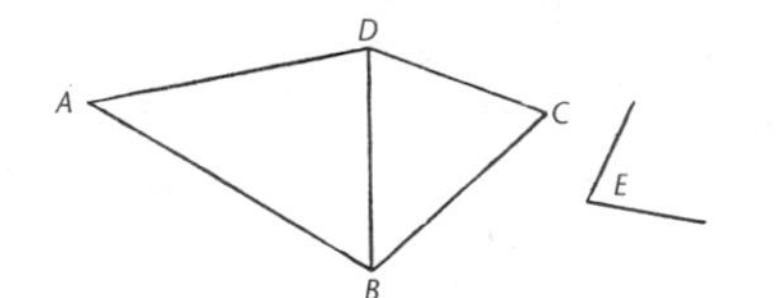

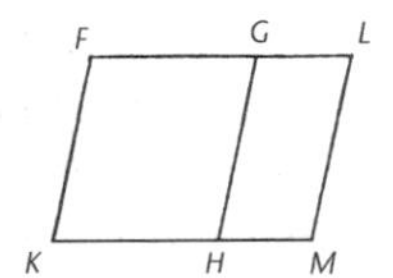

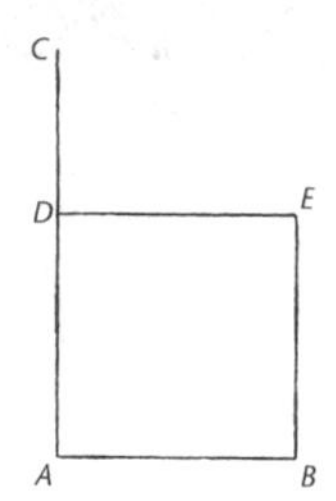

Proposition 46

On a given straight line to describe a square.

Proposition 47

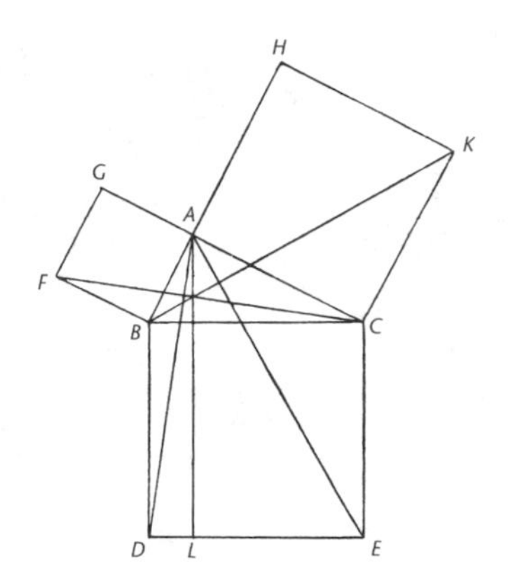

In right-angled triangles the square on the side subtending the right angle is equal to the squares on the sides containing the right angle.

Proposition 48

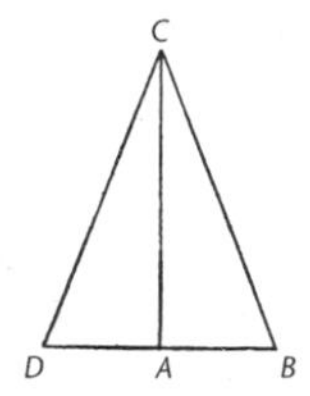

If in a triangle the square on one of the sides be equal to the squares on the remaining two sides of the triangle, the angle contained by the remaining two sides of the triangle is right.

Book II

Definitions

1. Any rectangular parallelogram is said to be *contained* by the two straight lines containing the right angle.
2. And in any parallelogrammic area let any one whatever of the parallelograms about its diameter with the two complements be called a *gnomon*.

Proposition 1

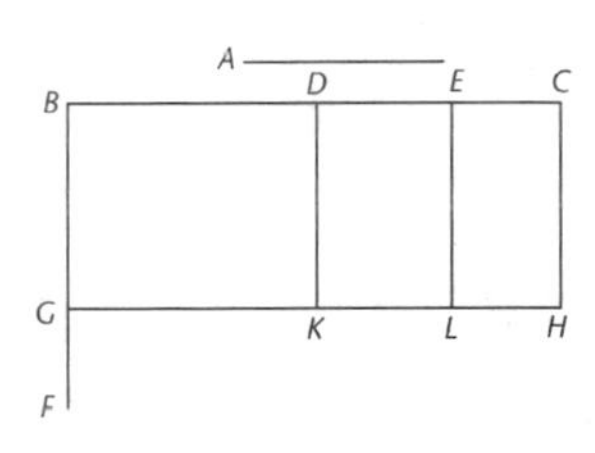

If there be two straight lines, and one of them be cut into any number of segments whatever, the rectangle contained by the two straight lines is equal to the rectangles contained by the uncut straight line and each of the segments.

Proposition 2

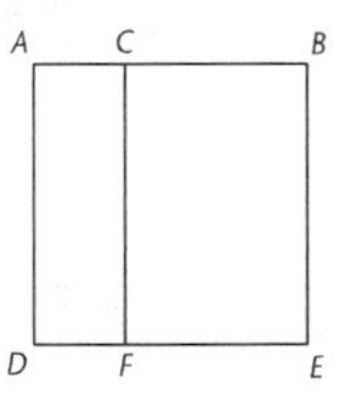

If a straight line be cut at random, the rectangles contained by the whole and both of the segments are equal to the square on the whole.

Proposition 3

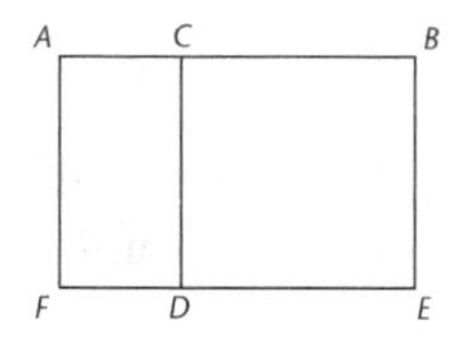

If a straight line be cut at random, the rectangle contained by the whole and one of the segments is equal to the rectangle contained by the segments and the square on the aforesaid segment.

Proposition 4

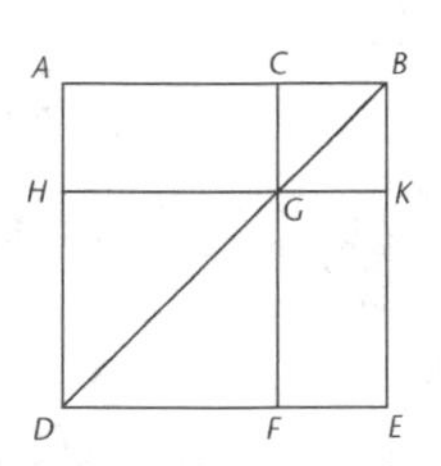

If a straight line be cut at random, the square on the whole is equal to the squares on the segments and twice the rectangle contained by the segments.

Proposition 5

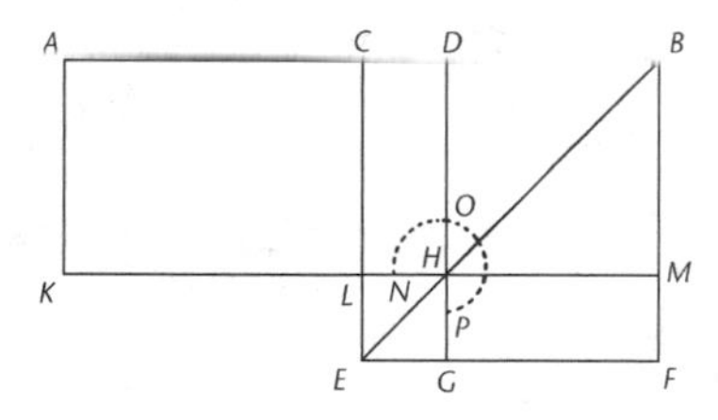

If a straight line be cut into equal and unequal segments, the rectangle contained by the unequal segments of the whole together with the square on the straight line between the points of section is equal to the square on the half.

Proposition 6

If a straight line be bisected and a straight line be added to it in a straight line, the rectangle contained by the whole with the added straight line and the added straight line together with the square on the half is equal to the square on the straight line made up of the half and the added straight line.

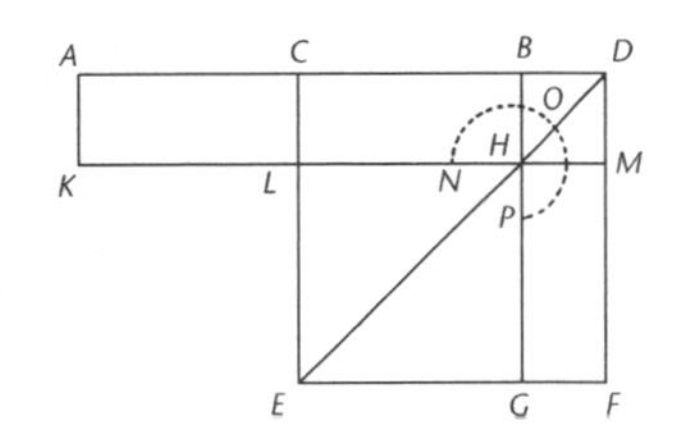

Proposition 7

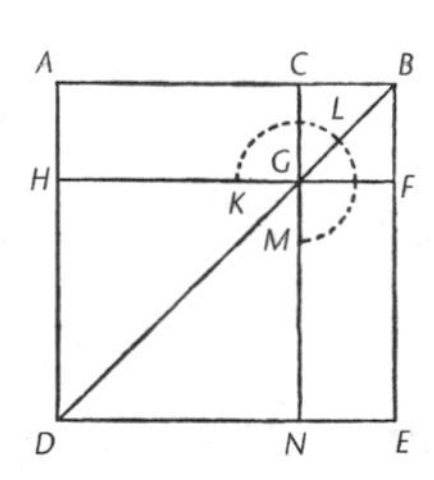

If a straight line be cut at random, the square on the whole and that on one of the segments both together are equal to twice the rectangle contained by the whole and the said segment and the square on the remaining segment.

Proposition 8

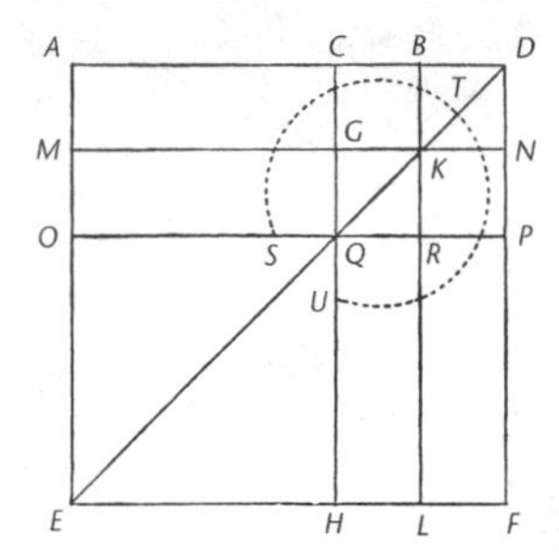

If a straight line be cut at random, four times the rectangle contained by the whole and one of the segments together with the square on the remaining segment is equal to the square described on the whole and the aforesaid segment as on one straight line.

Proposition 9

If a straight line be cut into equal and unequal segments, the squares on the unequal segments of the whole are double of the square on the half and of the square on the straight line between the points of section.

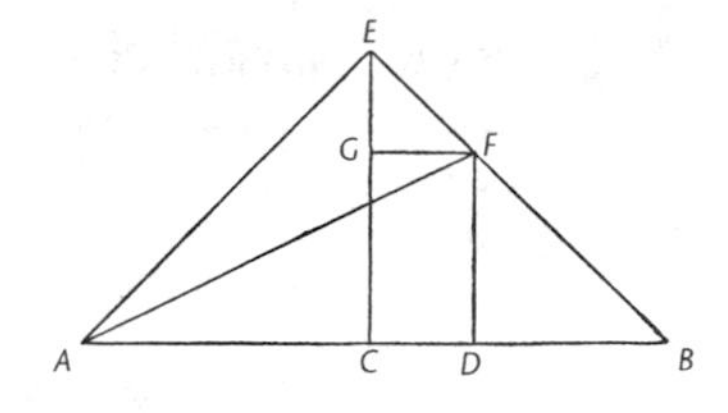

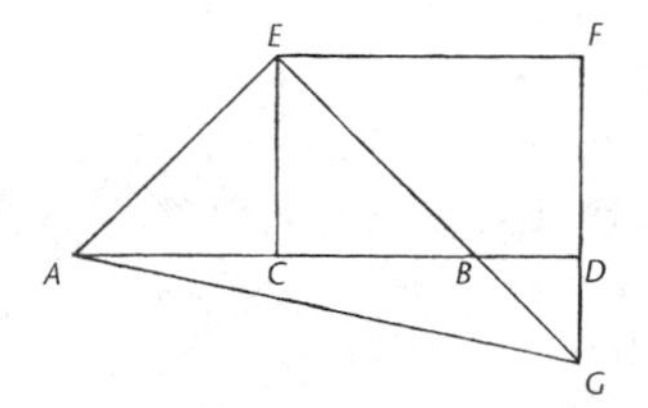

Proposition 10

If a straight line be bisected, and a straight line be added to it in a straight line, the square on the whole with the added straight line and the square on the added straight line both together are double of the square on the half and of the square described on the straight line made up of the half and the added straight line as on one straight line.

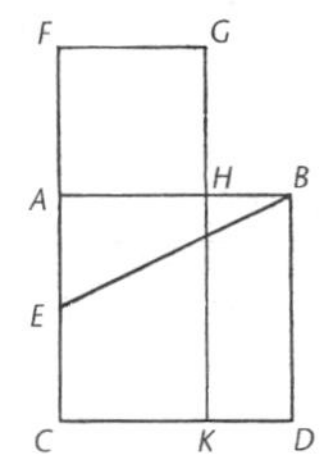

Proposition 11

To cut a given straight line so that the rectangle contained by the whole and one of the segments is equal to the square on the remaining segment.

Proposition 12

In obtuse-angled triangles the square on the side subtending the obtuse angle is greater than the squares on the sides containing the obtuse angle by twice the rectangle contained by one of the sides about the obtuse angle, namely that on which the perpendicular falls, and the straight line cut off outside by the perpendicular towards the obtuse angle.

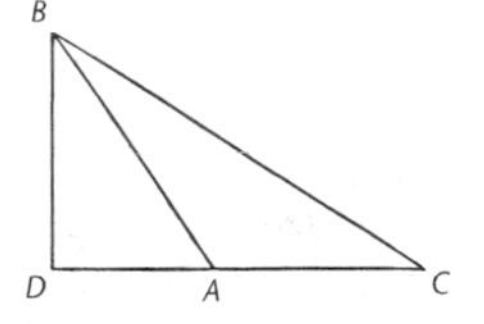

Proposition 13

In acute-angled triangles the square on the side subtending the acute angle is less than the squares on the sides containing the acute angle by twice the rectangle contained by one of the sides about the acute angle, namely that on which the perpendicular falls, and the straight line cut off within by the perpendicular towards the acute angle.

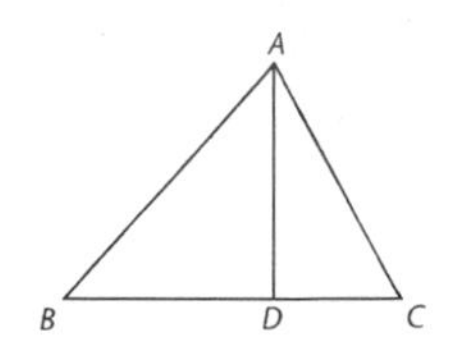

Proposition 14

To construct a square equal to a given rectilineal figure.

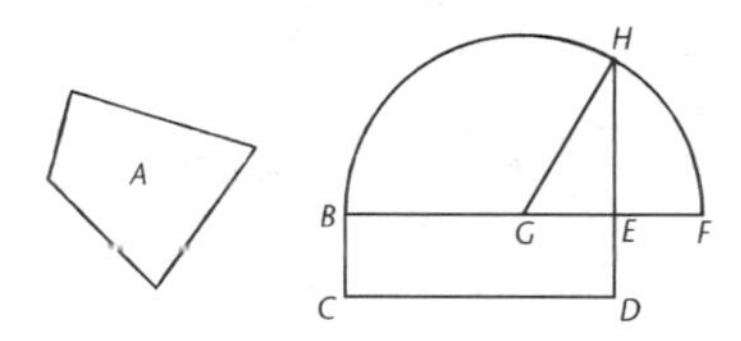

Book III

Definitions

1. *Equal circles* are those the diameters of which are equal, or the radii of which are equal.
2. A straight line is said to *touch a circle* which, meeting the circle and being produced, does not cut the circle.
3. *Circles* are said to *touch one another* which, meeting one another, do not cut one another.
4. In a circle straight lines are said *to be equally distant from the centre* when the perpendiculars drawn to them from the centre are equal.
5. And that straight line is said to be *at a greater distance* on which the greater perpendicular falls.
6. A *segment of a circle* is the figure contained by a straight line and a circumference of a circle.
7. An *angle of a segment* is that contained by a straight line and a circumference of a circle.
8. An *angle in a segment* is the angle which, when a point is taken on the circumference of the segment and straight lines are joined from it to the extremities of the straight line which is the *base of the segment,* is contained by the straight lines so joined.
9. And, when the straight lines containing the angle cut off a circumference, the angle is said to *stand upon* that circumference.

10. A *sector of a circle* is the figure which, when an angle is constructed at the centre of the circle, is contained by the straight lines containing the angle and the circumference cut off by them.
11. *Similar segments of circles* are those which admit equal angles, or in which the angles are equal to one another.

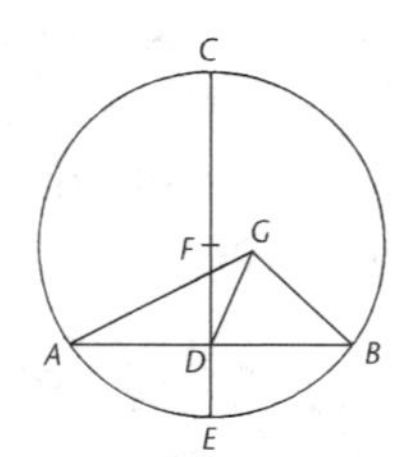

Proposition 1

To find the centre of a given circle.

PORISM. From this it is manifest that, if in a circle a straight line cut a straight line into two equal parts and at right angles, the centre of the circle is on the cutting straight line.

Proposition 2

If on the circumference of a circle two points be taken at random, the straight line joining the points will fall within the circle.

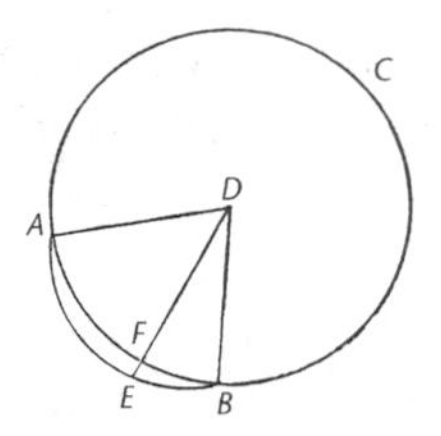

Proposition 3

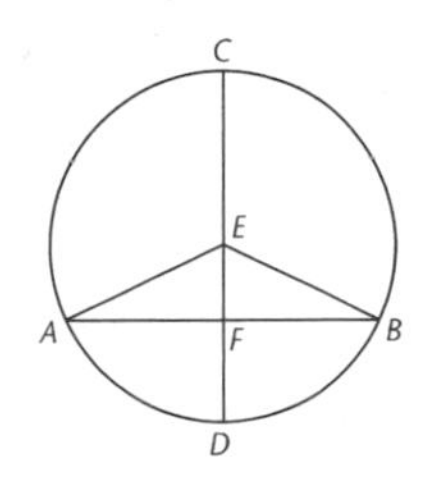

If in a circle a straight line through the centre bisect a straight line not through the centre, it also cuts it at right angles; and if it cut it at right angles, it also bisects it.

Proposition 4

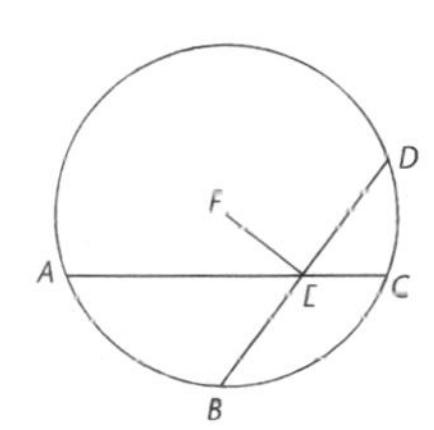

If in a circle two straight lines cut one another which are not through the centre, they do not bisect one another.

Proposition 5

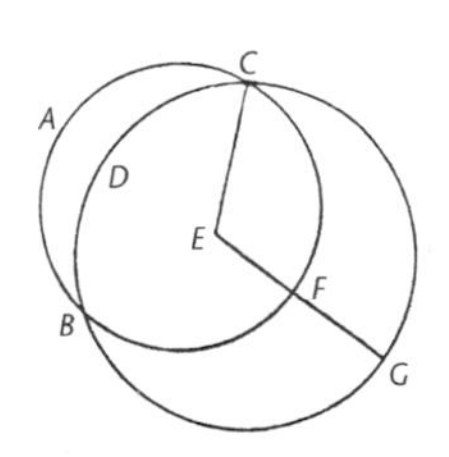

If two circles cut one another, they will not have the same centre.

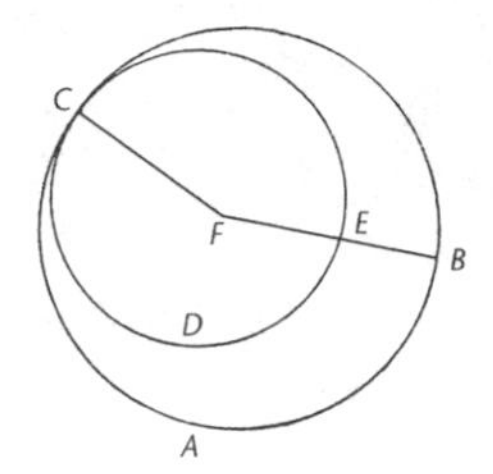

Proposition 6

If two circles touch one another, they will not have the same centre.

Proposition 7

If on the diameter of a circle a point be taken which is not the centre of the circle, and from the point straight lines fall upon the circle, that will be greatest on which the centre is, the remainder of the same diameter will be least, and of the rest the nearer to the straight line through the centre is always greater than the more remote, and only two equal straight lines will fall from the point on the circle, one on each side of the least straight line.

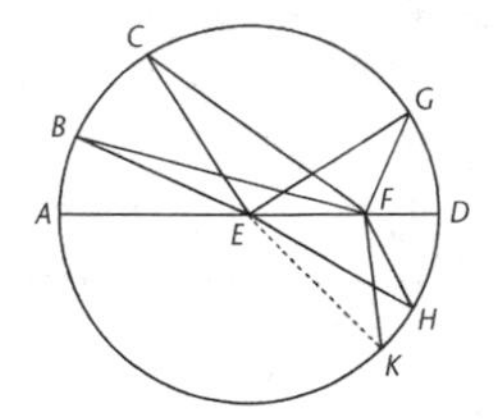

Proposition 8

If a point be taken outside a circle and from the point straight lines be drawn through to the circle, one of which is through the centre and the others are drawn at

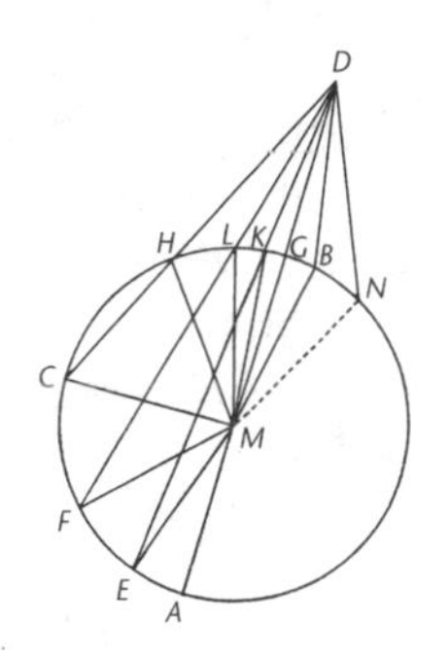

random, then, of the straight lines which fall on the concave circumference, that through the centre is greatest, while of the rest the nearer to that through the centre is always greater than the more remote, but, of the straight lines falling on the convex circumference, that between the point and the diameter is least, while of the rest the nearer to the least is always less than the more remote, and only two equal straight lines will fall on the circle from the point, one on each side of the least.

Proposition 9

If a point be taken within a circle, and more than two equal straight lines fall from the point on the circle, the point taken is the centre of the circle.

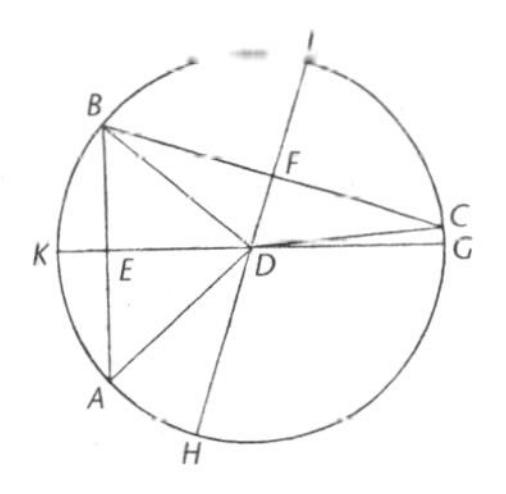

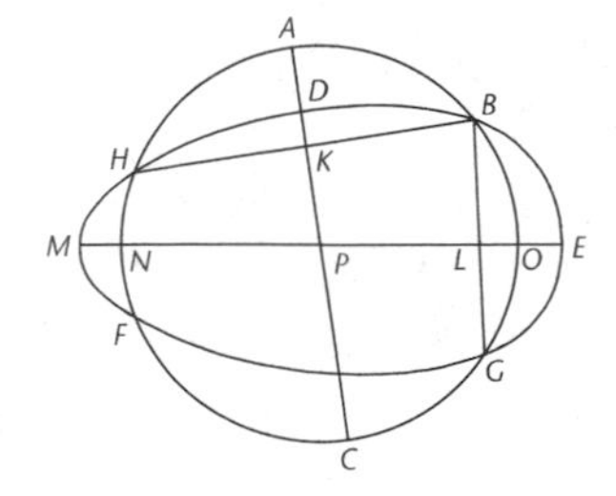

Proposition 10

A circle does not cut a circle at more points than two.

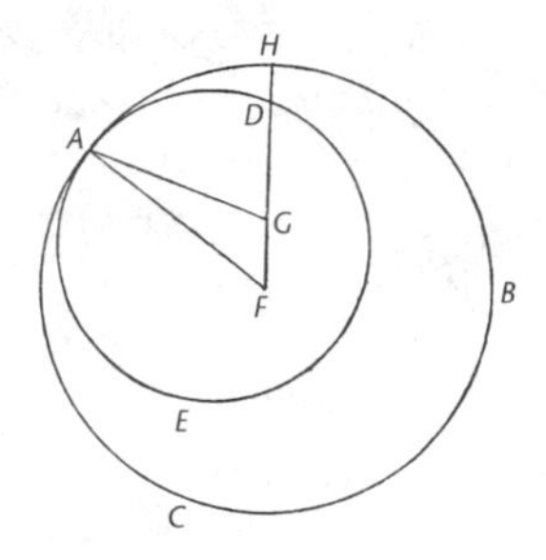

Proposition 11

If two circles touch one another internally, and their centres be taken, the straight line joining their centres, if it be also produced, will fall on the point of contact of the circles.

Proposition 12

If two circles touch one another externally, the straight line joining their centres will pass through the point of contact.

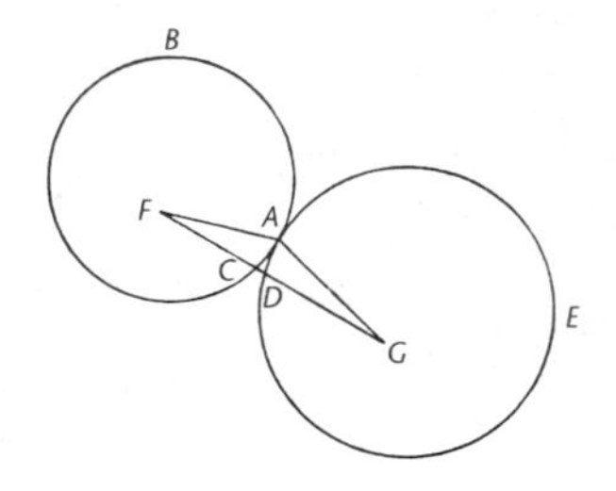

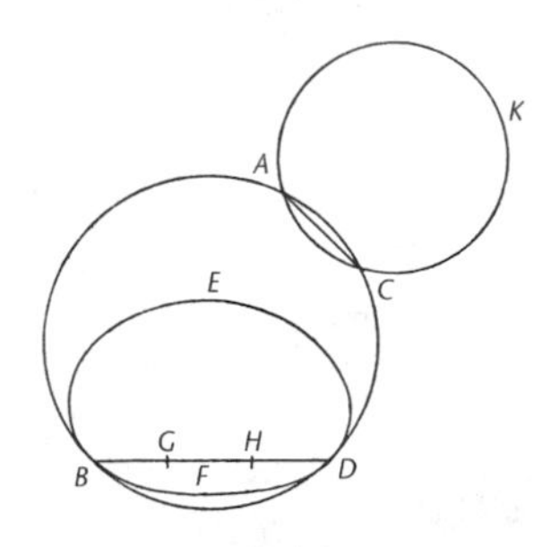

Proposition 13

A circle does not touch a circle at more points than one, whether it touch it internally or externally.

Proposition 14

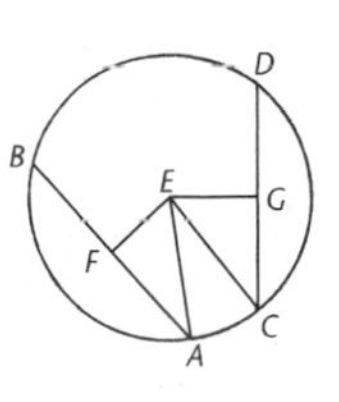

In a circle equal straight lines are equally distant from the centre, and those which are equally distant from the centre are equal to one another.

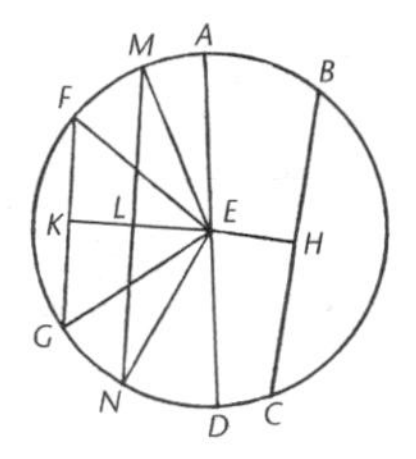

Proposition 15

Of straight lines in a circle the diameter is greatest, and of the rest the nearer to the centre is always greater than the more remote.

Proposition 16

The straight line drawn at right angles to the diameter of a circle from its extremity will fall outside the circle, and into the space between the straight line and the circumference another straight line cannot be interposed; further the angle of the semicircle is greater, and the remaining angle less, than any acute rectilineal angle.

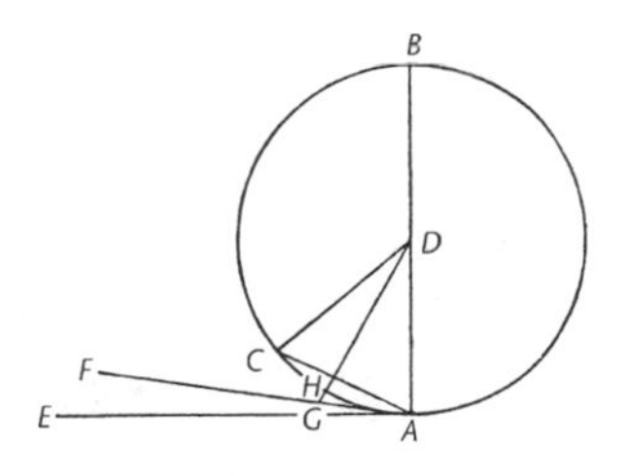

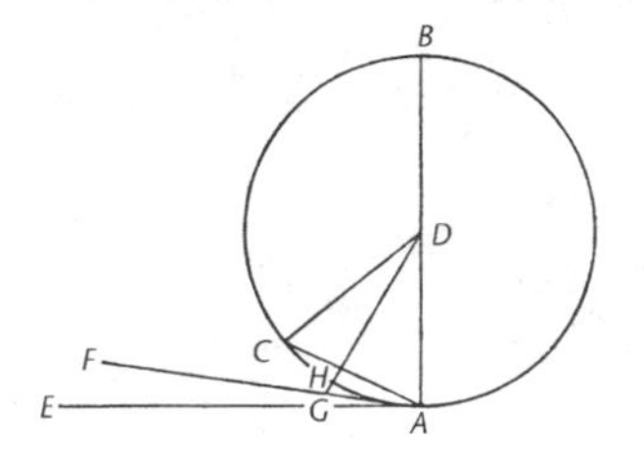

III.16 PORISM. From this it is manifest that the straight line drawn at right angles to the diameter of a circle from its extremity touches the circle.

Proposition 17

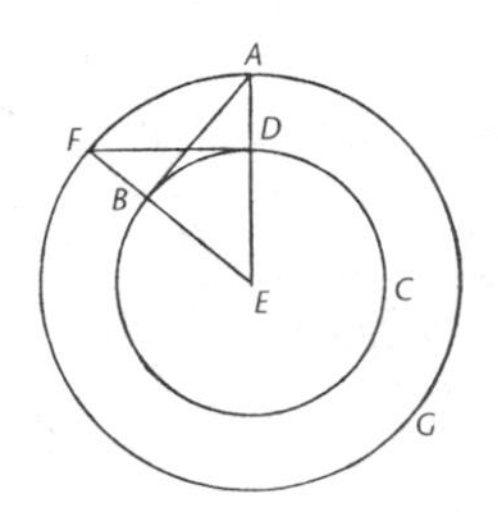

From a given point to draw a straight line touching a given circle.

Proposition 18

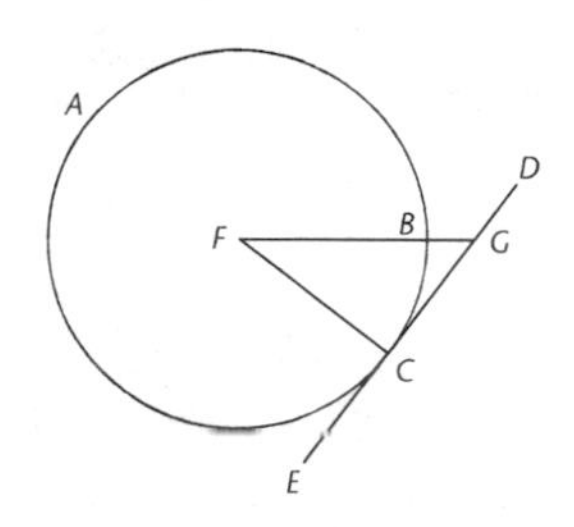

If a straight line touch a circle, and a straight line be joined from the centre to the point of contact, the straight line so joined will be perpendicular to the tangent.

Proposition 19

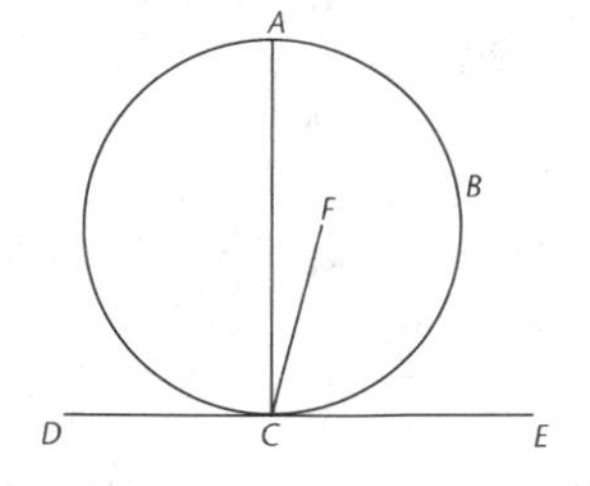

If a straight line touch a circle, and from the point of contact a straight line be drawn at right angles to the tangent, the centre of the circle will be on the straight line so drawn.

Proposition 20

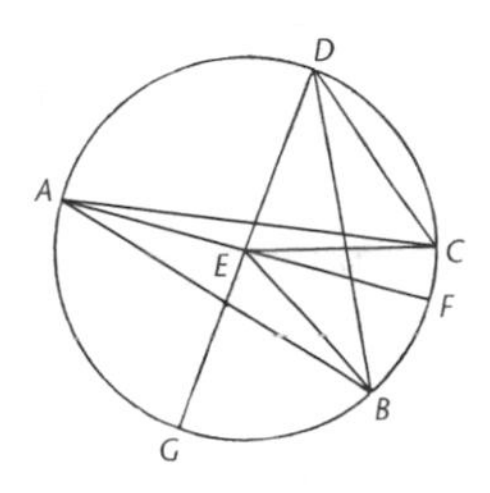

In a circle the angle at the centre is double of the angle at the circumference, when the angles have the same circumference as base.

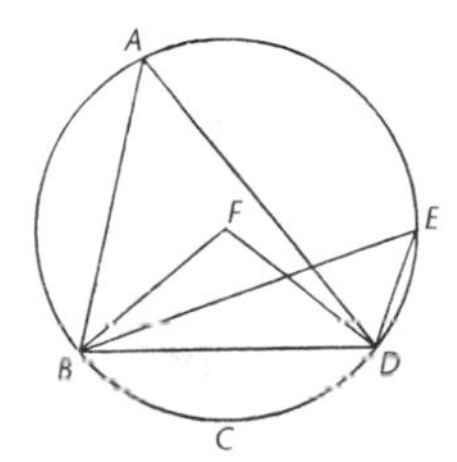

Proposition 21

In a circle the angles in the same segment are equal to one another.

Proposition 22

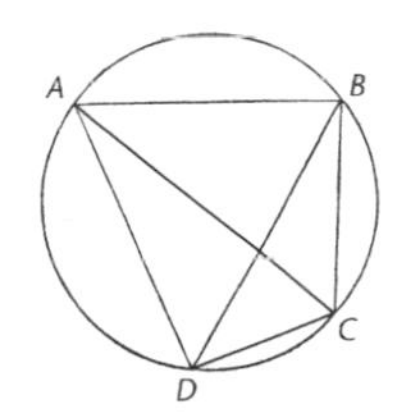

The opposite angles of quadrilaterals in circles are equal to two right angles.

Proposition 23

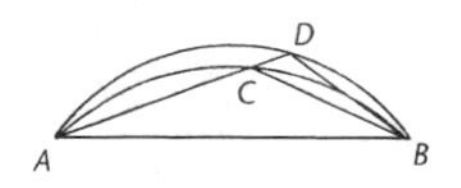

On the same straight line there cannot be constructed two similar and unequal segments of circles on the same side.

Proposition 24

Similar segments of circles on equal straight lines are equal to one another.

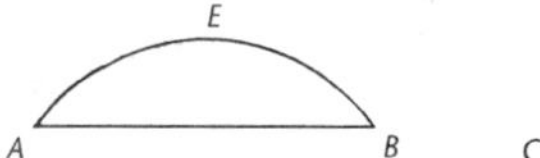

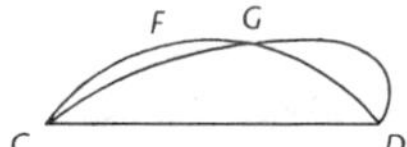

Proposition 25

Given a segment of a circle, to describe the complete circle of which it is a segment.

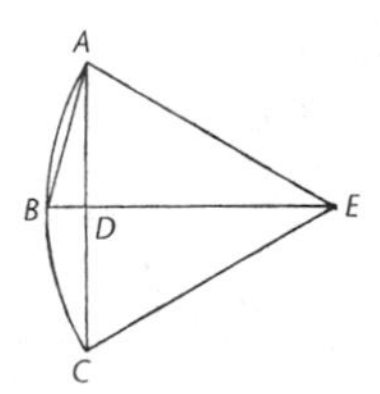

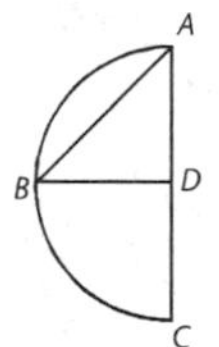

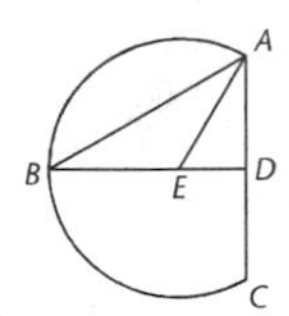

Proposition 26

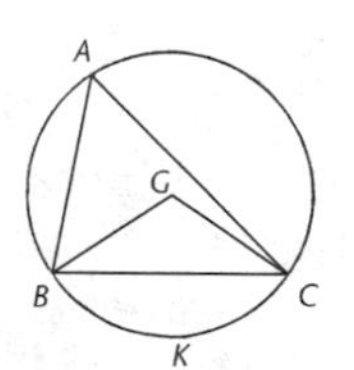

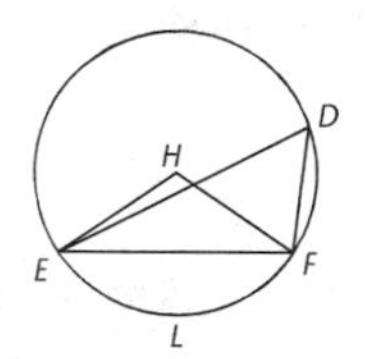

In equal circles equal angles stand on equal circumferences, whether they stand at the centres or at the circumferences.

Proposition 27

In equal circles angles standing on equal circumferences are equal to one another, whether they stand at the centres or at the circumferences.

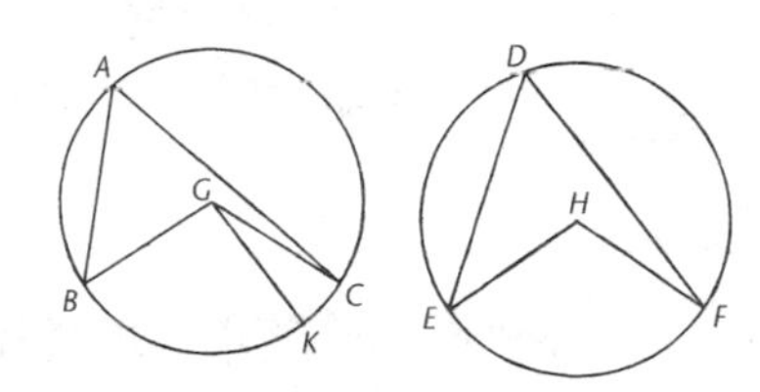

Proposition 28

In equal circles equal straight lines cut off equal circumferences, the greater equal to the greater and the less to the less.

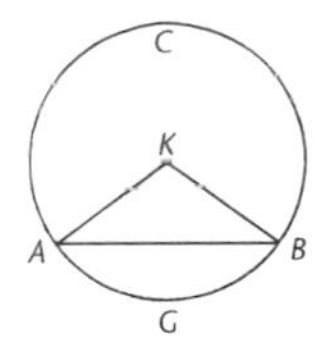

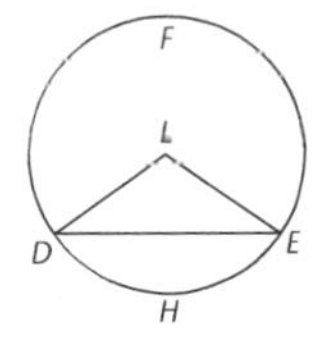

Proposition 29

In equal circles equal circumferences are subtended by equal straight lines.

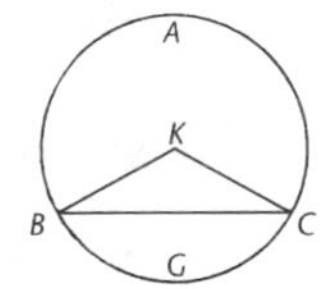

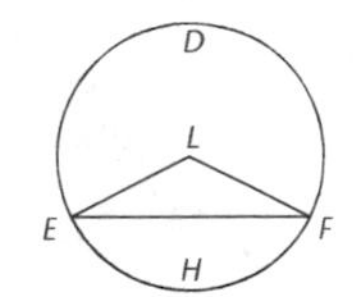

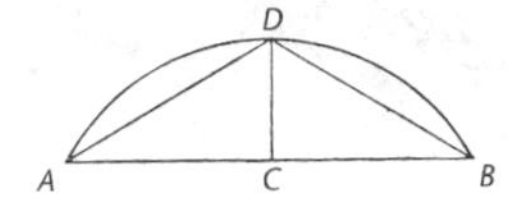

Proposition 30

To bisect a given circumference.

Proposition 31

In a circle the angle in the semicircle is right, that in a greater segment less than a right angle, and that in a less segment greater than a right angle; and further the angle of the greater segment is greater than a right angle, and the angle of the less segment less than a right angle.

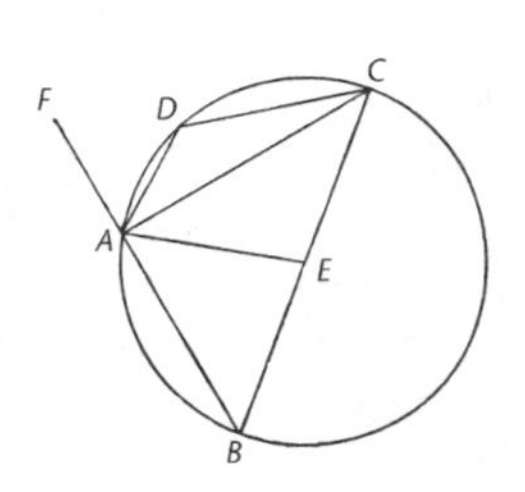

Proposition 32

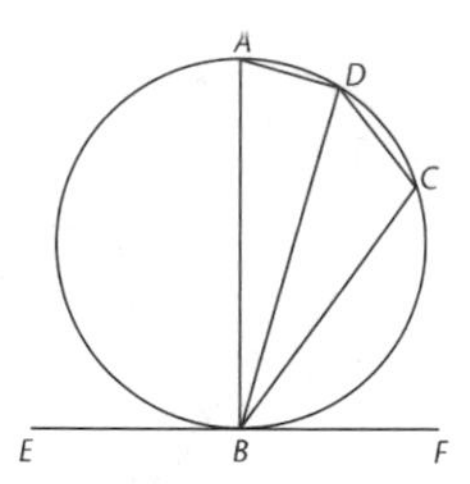

If a straight line touch a circle, and from the point of contact there be drawn across, in the circle, a straight line cutting the circle, the angles which it makes with the tangent will be equal to the angles in the alternate segments of the circle.

Proposition 33

On a given straight line to describe a segment of a circle admitting an angle equal to a given rectilineal angle.

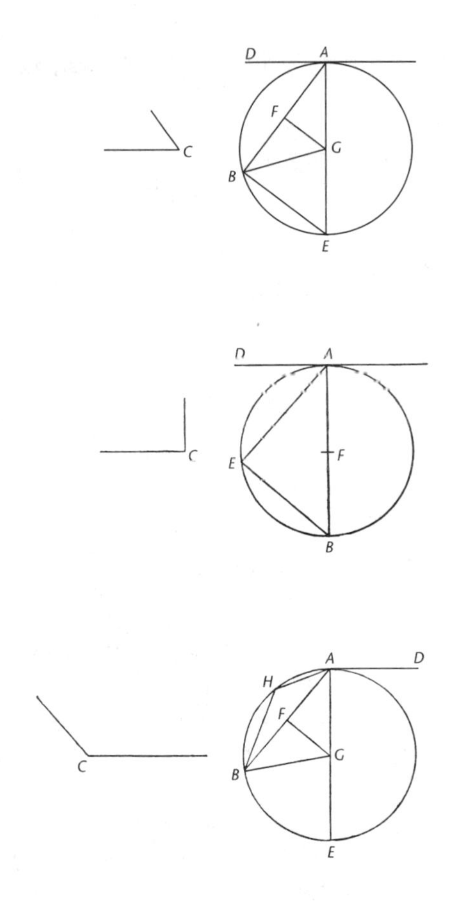

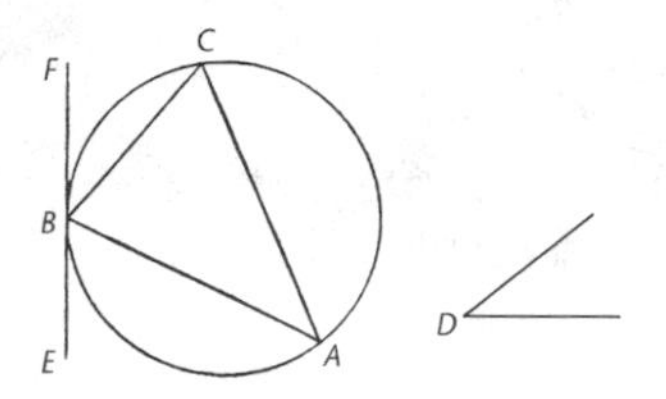

Proposition 34

From a given circle to cut off a segment admitting an angle equal to a given rectilineal angle.

Proposition 35

If in a circle two straight lines cut one another, the rectangle contained by the segments of the one is equal to the rectangle contained by the segments of the other.

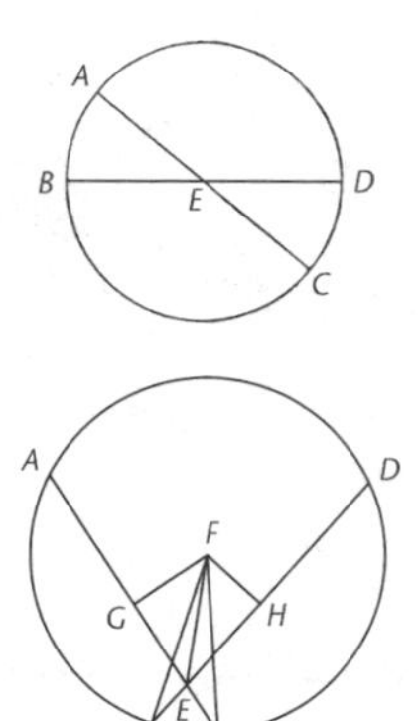

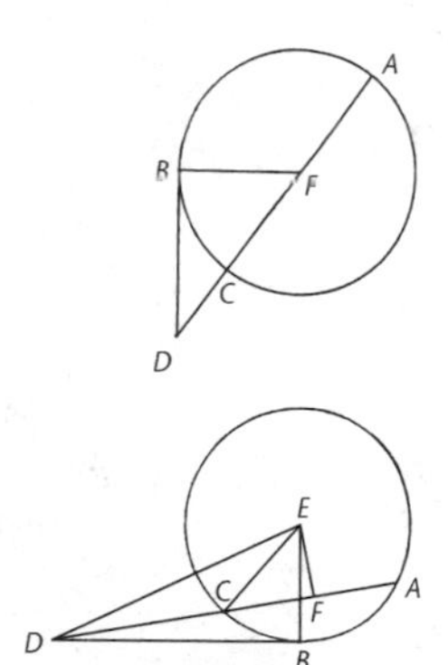

Proposition 36

If a point be taken outside a circle and from it there fall on the circle two straight lines, and if one of them cut the circle and the other touch it, the rectangle contained by the whole of the straight line which cuts the circle and the straight line intercepted on it outside between the point and the convex circumference will be equal to the square on the tangent.

Proposition 37

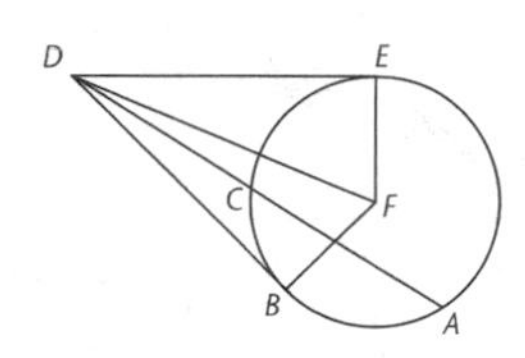

If a point be taken outside a circle and from the point there fall on the circle two straight lines, if one of them cut the circle, and the other fall on it, and if further the rectangle contained by the whole of the straight line which cuts the circle and the straight line intercepted on it outside between the point and the convex circumference be equal to the square on the straight line which falls on the circle, the straight line which falls on it will touch the circle.

Book IV

Definitions

1. A rectilineal figure is said to be *inscribed in a rectilineal figure* when the respective angles of the inscribed figure lie on the respective sides of that in which it is inscribed.
2. Similarly a figure is said to be *circumscribed about a figure* when the respective sides of the circumscribed figure pass through the respective angles of that about which it is circumscribed.
3. A rectilineal figure is said to be *inscribed in a circle* when each angle of the inscribed figure lies on the circumference of the circle.
4. A rectilineal figure is said to be *circumscribed about a circle,* when each side of the circumscribed figure touches the circumference of the circle.
5. Similarly a circle is said to be *inscribed in a figure* when the circumference of the circle touches each side of the figure in which it is inscribed.
6. A circle is said to be *circumscribed about a figure* when the circumference of the circle passes through each angle of the figure about which it is circumscribed.
7. A straight line is said to be *fitted into a circle* when its extremities are on the circumference of the circle.

Proposition 1

Into a given circle to fit a straight line equal to a given straight line which is not greater than the diameter of the circle.

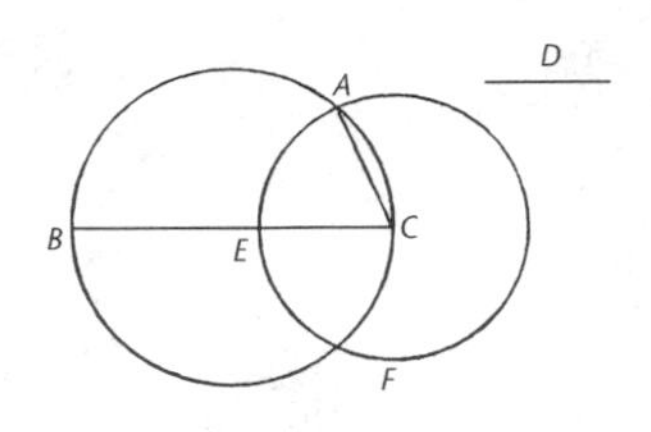

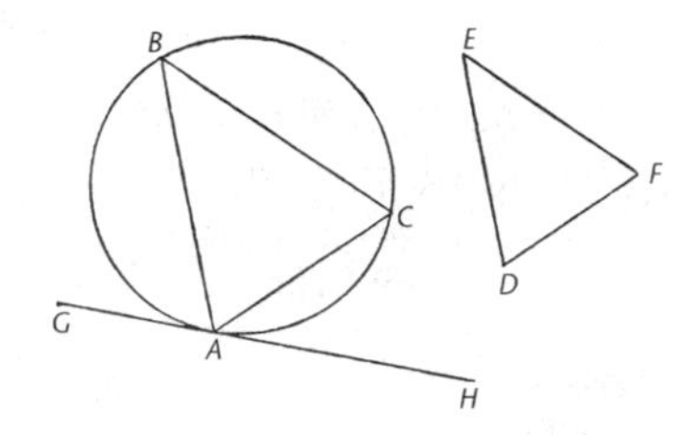

Proposition 2

In a given circle to inscribe a triangle equiangular with a given triangle.

Proposition 3

About a given circle to circumscribe a triangle equiangular with a given triangle.

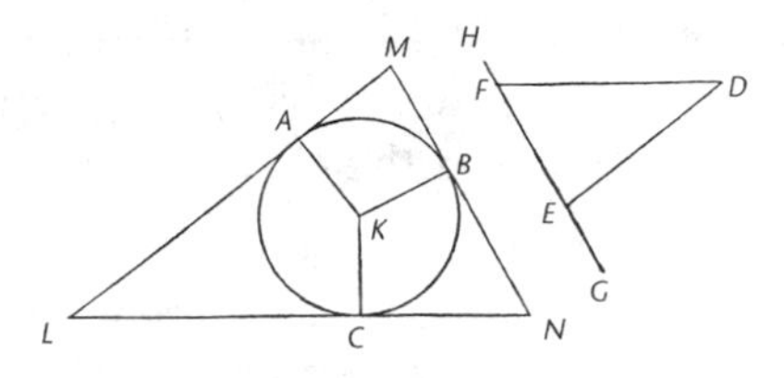

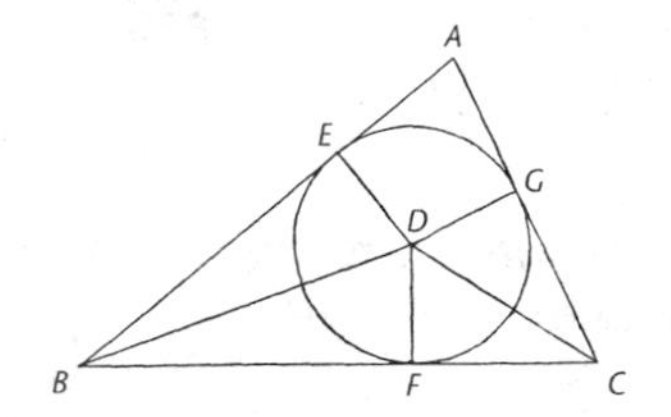

Proposition 4

In a given triangle to inscribe a circle.

Proposition 5

About a given triangle to circumscribe a circle.

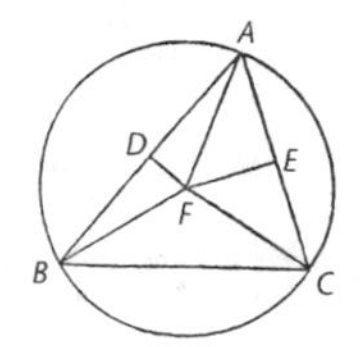

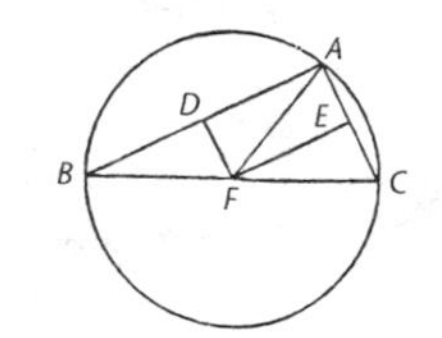

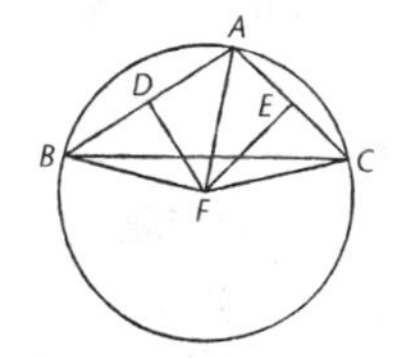

Proposition 6

In a given circle to inscribe a square.

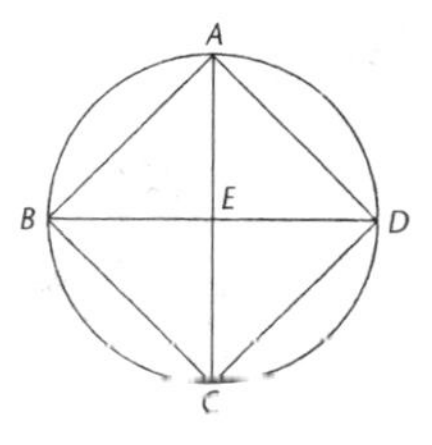

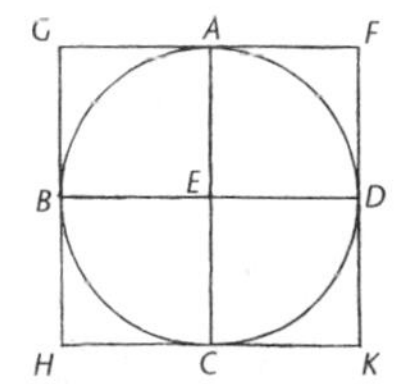

Proposition 7

About a given circle to circumscribe a square.

Proposition 8

In a given square to inscribe a circle.

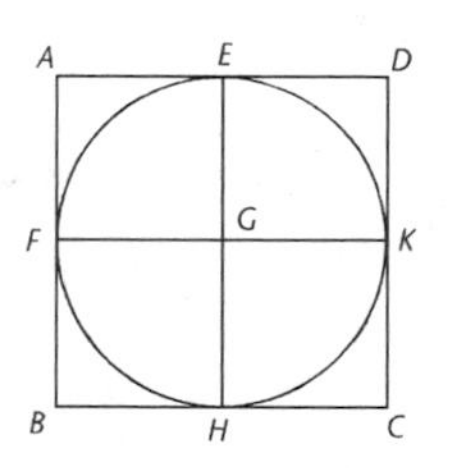

Proposition 9

About a given square to circumscribe a circle.

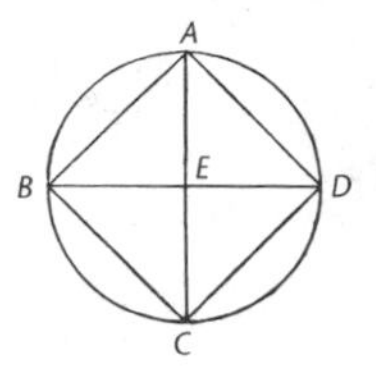

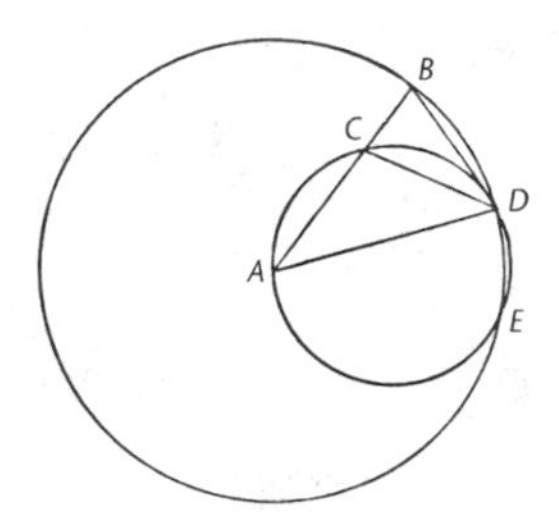

Proposition 10

To construct an isosceles triangle having each of the angles at the base double of the remaining one.

Proposition 11

In a given circle to inscribe an equilateral and equiangular pentagon.

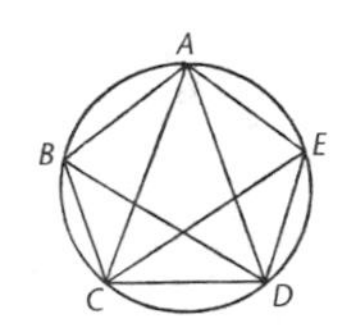

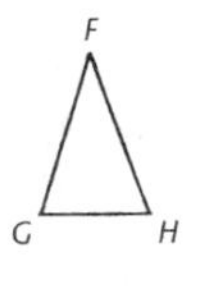

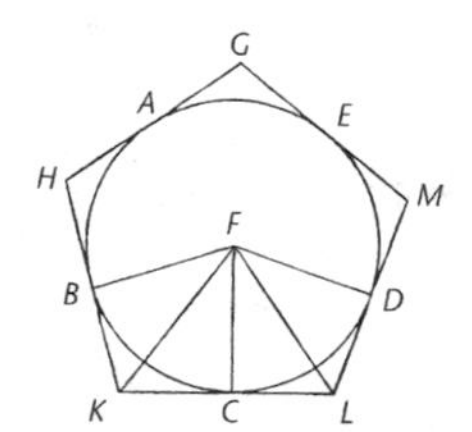

Proposition 12

About a given circle to circumscribe an equilateral and equiangular pentagon.

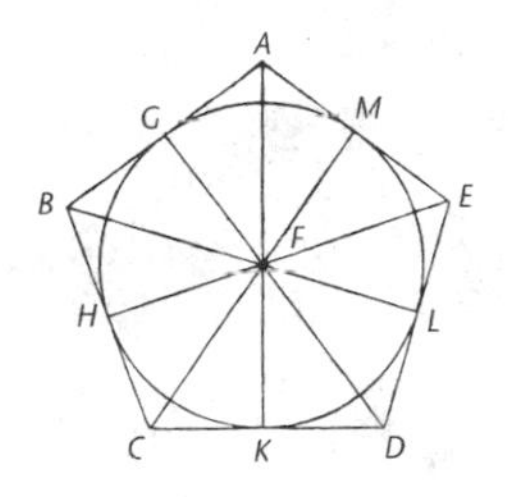

Proposition 13

In a given pentagon, which is equilateral and equiangular, to inscribe a circle.

Proposition 14

About a given pentagon, which is equilateral and equiangular, to circumscribe a circle.

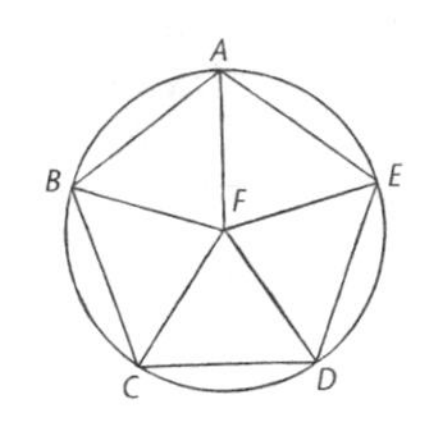

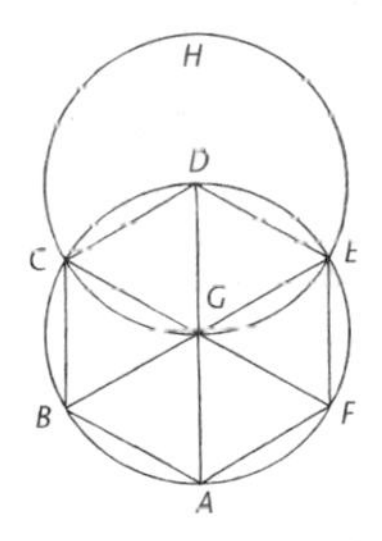

Proposition 15

In a given circle to inscribe an equilateral and equiangular hexagon.

Proposition 16

In a given circle to inscribe a fifteen-angled figure which shall be both equilateral and equiangular.

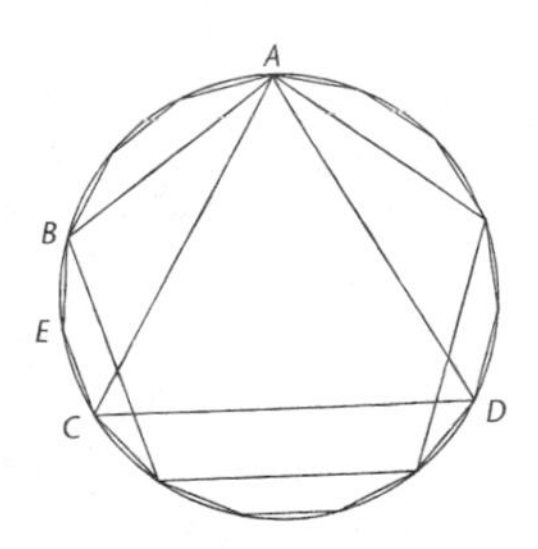

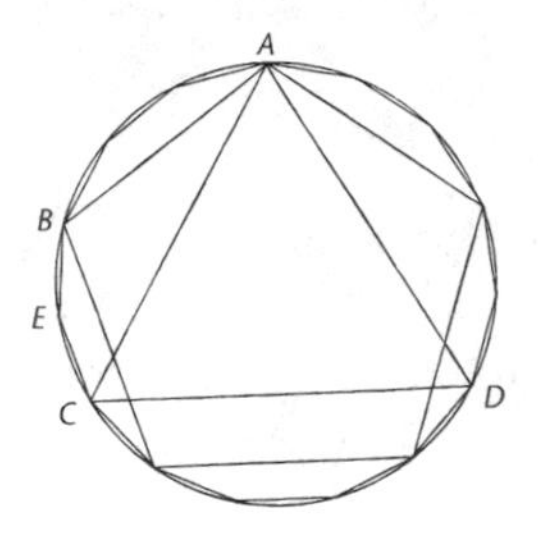

IV. 16 COROLLARY. And, in like manner as in the case of the pentagon, if through the points of division on the circle we draw tangents to the circle, there will be circumscribed about the circle a fifteen-angled figure which is equilateral and equiangular.

Book V

Definitions

1. A magnitude is a *part* of a magnitude, the less of the greater, when it measures the greater.
2. The greater is a *multiple* of the less when it is measured by the less.
3. A *ratio* is a sort of relation in respect of size between two magnitudes of the same kind.
4. Magnitudes are said to *have a ratio* to one another which are capable, when multiplied, of exceeding one another.
5. Magnitudes are said to *be in the same ratio,* the first to the second and the third to the fourth, when, if any equimultiples whatever be taken of the first and third, and any equimultiples whatever of the second and fourth, the former equimultiples alike exceed, are alike equal to, or alike fall short of, the latter equimultiples respectively taken in corresponding order.
6. Let magnitudes which have the same ratio be called *proportional.*
7. When, of the equimultiples, the multiple of the first magnitude exceeds the multiple of the second, but the multiple of the third does not exceed the multiple of the fourth, then the first is said to *have a greater ratio* to the second than the third has to the fourth.

8. A proportion in three terms is the least possible.

9. When three magnitudes are proportional, the first is said to have to the third the *duplicate ratio* of that which it has to the second.

10. When four magnitudes are [continuously] proportional, the first is said to have to the fourth the *triplicate ratio* of that which it has to the second, and so on continually, whatever be the proportion.[1]

11. The term *corresponding magnitudes* is used of antecedents in relation to antecedents, and of consequents in relation to consequents.

12. *Alternate ratio* means taking the antecedent in relation to the antecedent and the consequent in relation to the consequent.[2]

13. *Inverse ratio* means taking the consequent as antecedent in relation to the antecedent as consequent.[3]

14. *Composition of a ratio* means taking the antecedent together with the consequent as one in relation to

---

1. If four magnitudes a, b, c, d are such that $a:b :: b:c :: c:d$, the ratio $a:d$ is said to be triplicate of the ratio $a:b$. —Ed.

2. If $a:b :: c:d$, the proportion $a:c :: b:d$ is obtained "alternately" (*alternando*), as proved in Proposition V. 16. —Ed.

3. If $a:b :: c:d$, the proportion $b:a :: d:c$ is obtained "inversely" (*invertendo*), as proved in Proposition V. 7, Porism. —Ed.

the consequent by itself.[4]

15. *Separation of a ratio* means taking the excess by which the antecedent exceeds the consequent in relation to the consequent by itself.[5]

16. *Conversion of a ratio* means taking the antecedent in relation to the excess by which the antecedent exceeds the consequent.[6]

17. A ratio *ex aequali* arises when, there being several magnitudes and another set equal to them in multitude which taken two and two are in the same proportion, as the first is to the last among the first magnitudes, so is the first to the last among the second magnitudes;

    Or, in other words, it means taking the extreme terms by virtue of the removal of the intermediate terms.[7]

---

4. If $a:b :: c:d$, the proportion $a+b:b :: c+d:d$ is obtained by composition (*componendo*). See Propositions V. 17 and V. 18. —Ed.

5. If $a:b :: c:d$, the proportion $a-b:b :: c-d:d$ is obtained by separation (*separando*). See Propositions V. 17 and V. 18. —Ed.

6. If $a:b :: c:d$, the proportion $a:a-b :: c:c-d$ is obtained by conversion (*convertendo*). See Proposition V. 19. —Ed.

7. If $a:b :: d:e$
and also $b:c :: e:f$
then the proportion $a:c :: d:f$
is obtained *ex aequali*. Note the similar disposition of middle terms b and e. The example can be extended to any number of pairs of magnitudes, as proved in Proposition V. 22. —Ed.

18. A *perturbed proportion* arises when, there being three magnitudes and another set equal to them in multitude, as antecedent is to consequent among the first magnitudes, so is antecedent to consequent among the second magnitudes, while, as the consequent is to a third among the first magnitudes, so is a third to the antecedent among the second magnitudes.[8]

Proposition 1

If there be any number of magnitudes whatever which are, respectively, equimultiples of any magnitudes equal in multitude, then, whatever multiple one of the magnitudes is of one, that multiple also will all be of all.

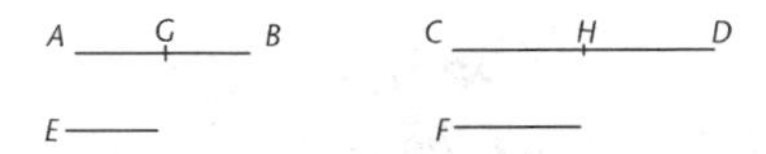

8. If $a:b :: e:f$
and also $b:c :: d:e$
then the proportion is said to be perturbed, since the middle terms b and e are oppositely disposed. The proportion $a:c :: d:f$ is obtained *ex aequali* in this case too, as proved in Proposition V. 23. —Ed.

Proposition 2

If a first magnitude be the same multiple of a second that a third is of a fourth, and a fifth also be the same multiple of the second that a sixth is of the fourth, the sum of the first and fifth will also be the same multiple of the second that the sum of the third and sixth is of the fourth.

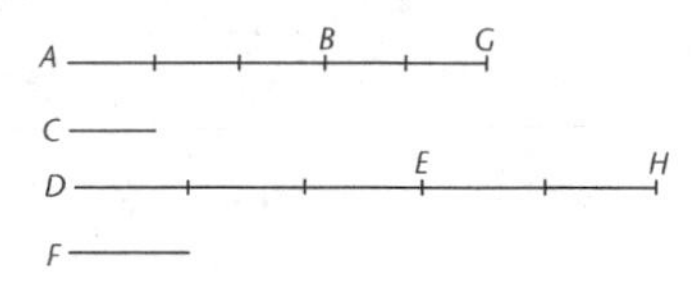

Proposition 3

If a first magnitude be the same multiple of a second that a third is of a fourth, and if equimultiples be taken of the first and third, then also *ex aequali* the magnitudes taken will be equimultiples respectively, the one of the second, and the other of the fourth.

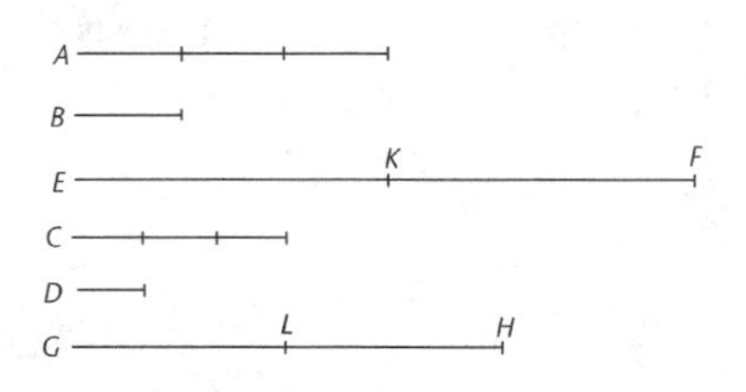

Proposition 4

If a first magnitude have to a second the same ratio as a third to a fourth, any equimultiples whatever of the first and third will also have the same ratio to any equimultiples whatever of the second and fourth respectively, taken in corresponding order.

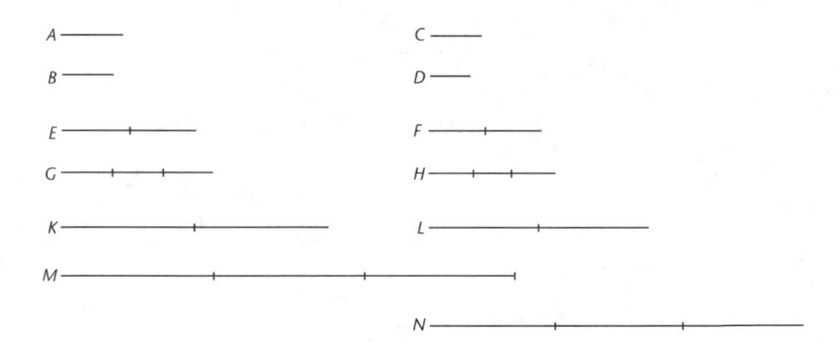

Proposition 5

If a magnitude be the same multiple of a magnitude that a part subtracted is of a part subtracted, the remainder will also be the same multiple of the remainder that the whole is of the whole.

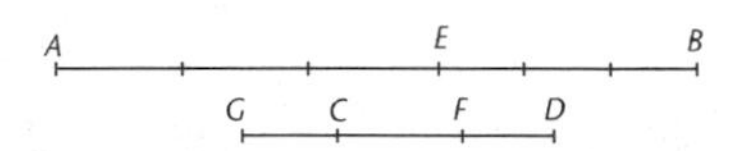

Proposition 6

If two magnitudes be equimultiples of two magnitudes, and any magnitudes subtracted from them be equimultiples of the same, the remainders also are either equal to the same or equimultiples of them.

Proposition 7

Equal magnitudes have to the same the same ratio, as also has the same to equal magnitudes.

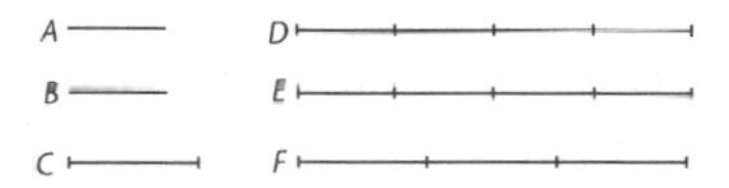

PORISM. From this it is manifest that, if any magnitudes are proportional, they will also be proportional inversely.

Proposition 8

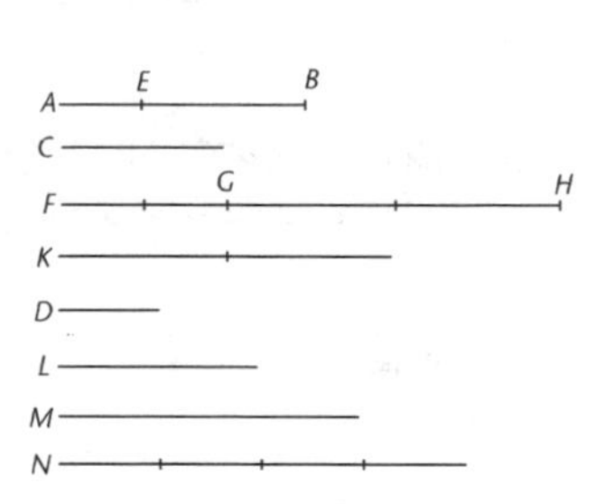

Of unequal magnitudes, the greater has to the same a greater ratio than the less has; and the same has to the less a greater ratio than it has to the greater.

Proposition 9

Magnitudes which have the same ratio to the same are equal to one another; and magnitudes to which the same has the same ratio are equal.

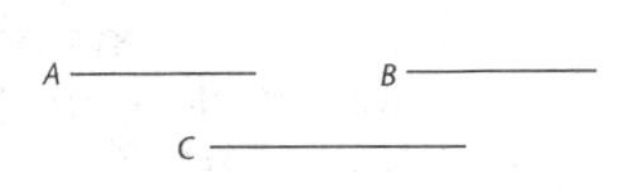

Proposition 10

Of magnitudes which have a ratio to the same, that which has a greater ratio is greater; and that to which the same has a greater ratio is less.

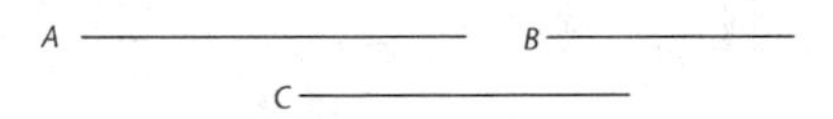

Proposition 11

Ratios which are the same with the same ratio are also the same with one another.

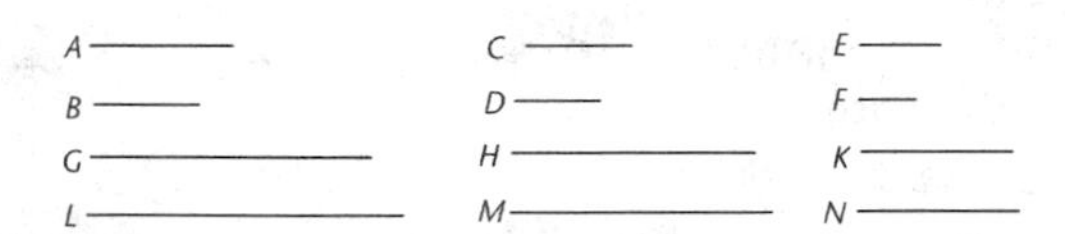

Proposition 12

If any number of magnitudes be proportional, as one of the antecedents is to one of the consequents, so will all the antecedents be to all the consequents.

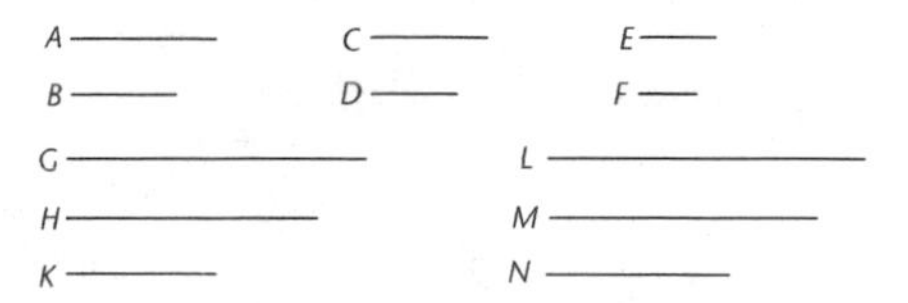

Proposition 13

If a first magnitude have to a second the same ratio as a third to a fourth, and the third have to the fourth a greater ratio than a fifth has to a sixth, the first will also have to the second a greater ratio than the fifth to the sixth.

Proposition 14

If a first magnitude have to a second the same ratio as a third has to a fourth, and the first be greater than the third, the second will also be greater than the fourth; if equal, equal; and if less, less.

Proposition 15

Parts have the same ratio as the same multiples of them taken in corresponding order.

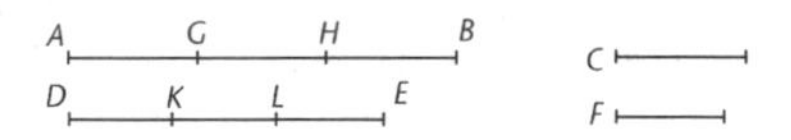

Proposition 16

If four magnitudes be proportional, they will also be proportional alternately.

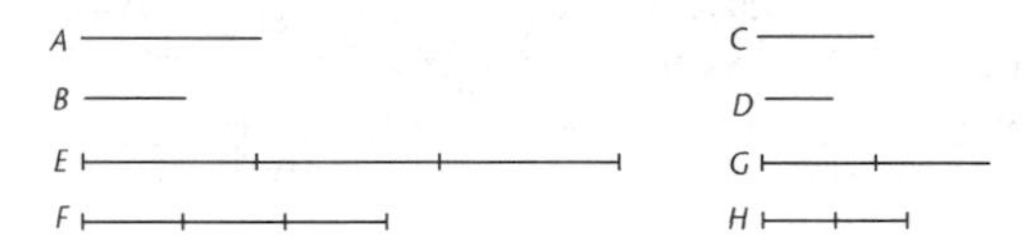

Proposition 17

If magnitudes be proportional *componendo*, they will also be proportional *separando*.

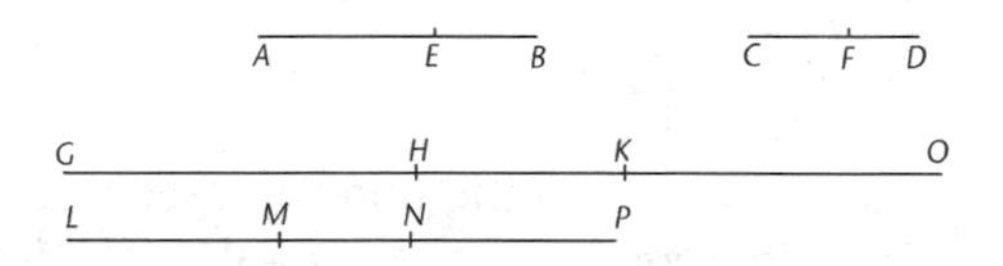

Proposition 18

If magnitudes be proportional *separando*, they will also be proportional *componendo*.

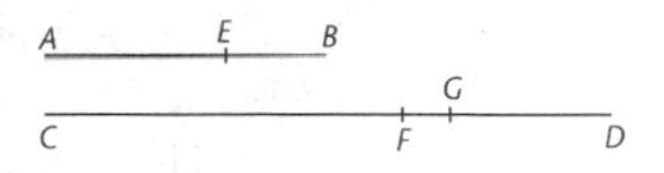

Proposition 19

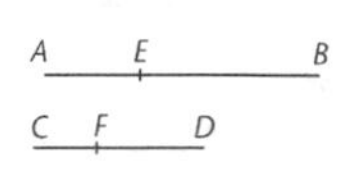

If, as a whole is to a whole, so is a part subtracted to a part subtracted, the remainder will also be to the remainder as whole to whole.

⟨**PORISM.** From this it is manifest that, if magnitudes be proportional *componendo* [V. Def. 14], they will also be pro portional *convertendo* [V. Def. 16].⟩

Proposition 20

If there be three magnitudes, and others equal to them in multitude, which taken two and two are in the same ratio, and if *ex aequali* the first be greater than the third, the fourth will also be greater than the sixth; if equal, equal; and, if less, less.

Proposition 21

If there be three magnitudes, and others equal to them in multitude, which taken two and two together are in the same ratio, and the proportion of them be perturbed, then, if *ex aequali* the first magnitude is greater than the third, the fourth will also be greater than the sixth; if equal, equal; and if less, less.

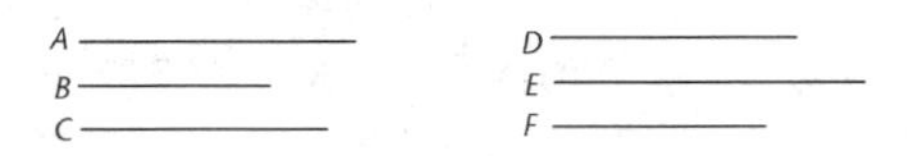

Proposition 22

If there be any number of magnitudes whatever, and others equal to them in multitude, which taken two and two together are in the same ratio, they will also be in the same ratio *ex aequali*.

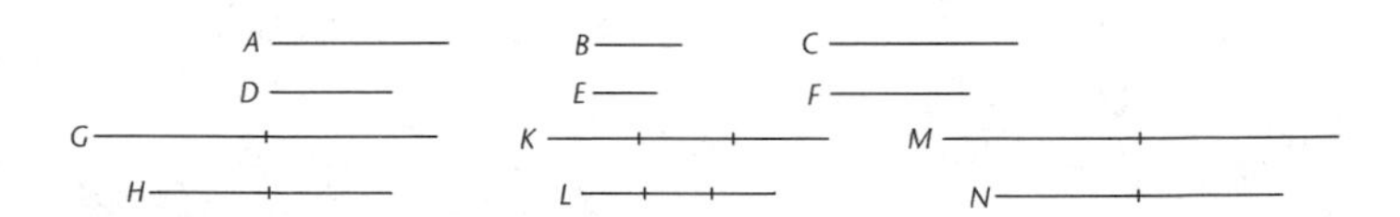

Proposition 23

If there be three magnitudes, and others equal to them in multitude, which taken two and two together are in the same ratio, and the proportion of them be perturbed, they will also be in the same ratio *ex aequali*.

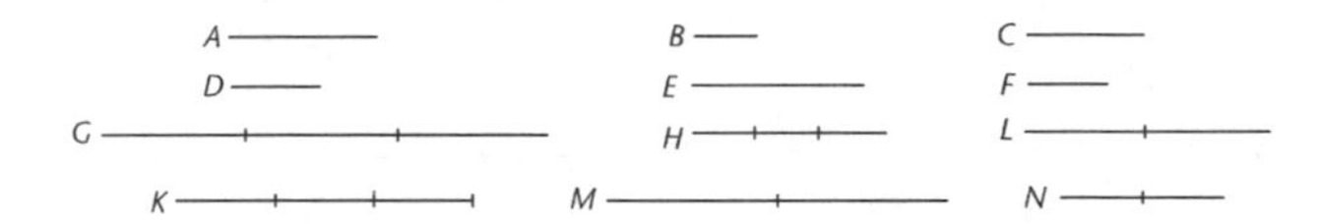

Proposition 24

If a first magnitude have to a second the same ratio as a third has to a fourth, and also a fifth have to the second the same ratio as a sixth to the fourth, the first and fifth added together will have to the second the same ratio as the third and sixth have to the fourth.

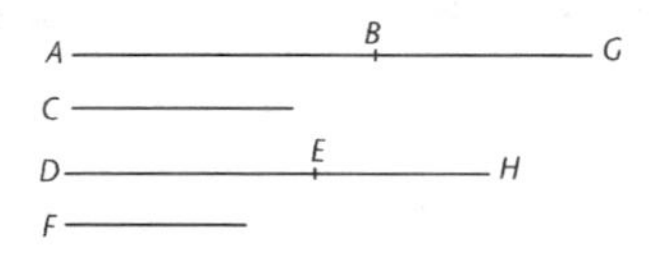

Proposition 25

If four magnitudes be proportional, the greatest and the least are greater than the remaining two.

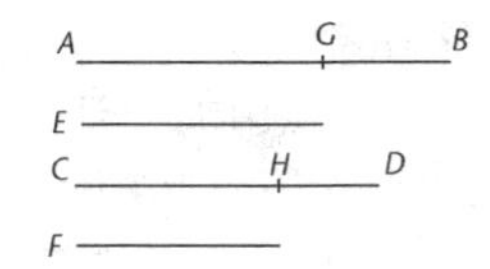

Book VI

Definitions

1. *Similar rectilineal figures* are such as have their angles severally equal and the sides about the equal angles proportional.

⟨2. *Reciprocally related* figures.⟩[1]

3. A straight line is said to have been *cut in extreme and mean ratio* when, as the whole line is to the greater segment, so is the greater to the less.

4. The *height* of any figure is the perpendicular drawn from the vertex to the base.

Proposition 1

Triangles and parallelograms which are under the same height are to one another as their bases.

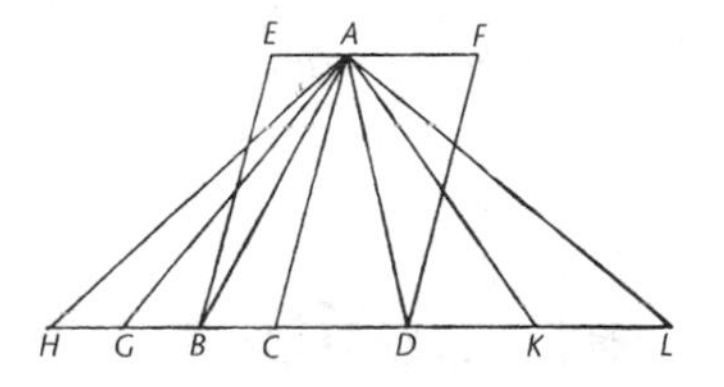

1. Definition 2 is regarded as spurious. It is never used by Euclid. According to Heath, its Greek text conveys no intelligible meaning, and Heath does not translate it. —Ed.

Proposition 2

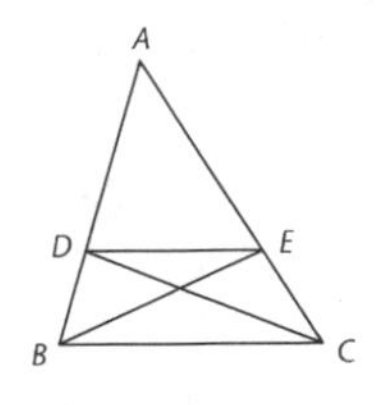

If a straight line be drawn parallel to one of the sides of a triangle, it will cut the sides of the triangle proportionally; and, if the sides of the triangle be cut proportionally, the line joining the points of section will be parallel to the remaining side of the triangle.

Proposition 3

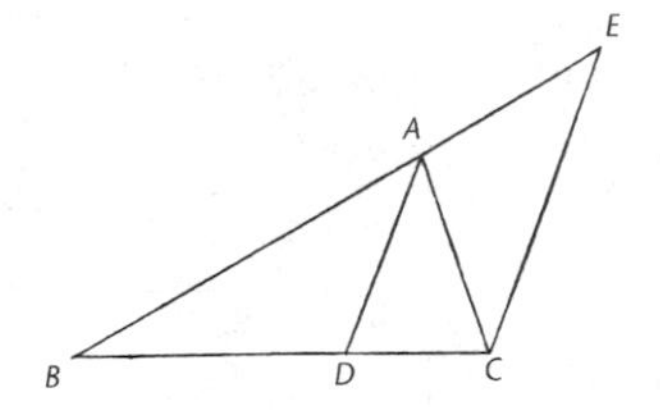

If an angle of a triangle be bisected and the straight line cutting the angle cut the base also, the segments of the base will have the same ratio as the remaining sides of the triangle; and, if the segments of the base have the same ratio as the remaining sides of the triangle, the straight line joined from the vertex to the point of section will bisect the angle of the triangle.

Proposition 4

In equiangular triangles the sides about the equal angles are proportional, and those are corresponding sides which subtend the equal angles.

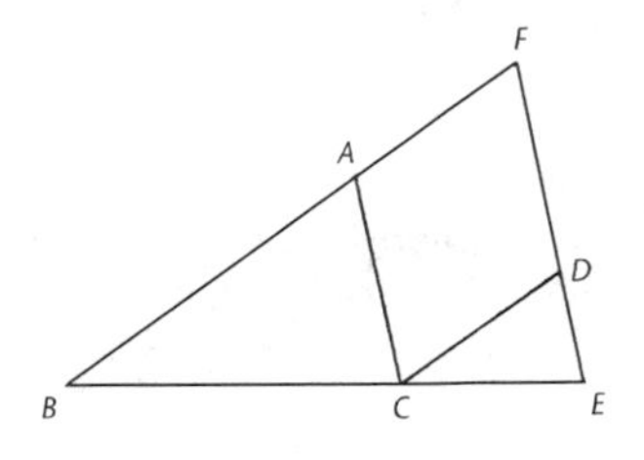

Proposition 5

If two triangles have their sides proportional, the triangles will be equiangular and will have those angles equal which the corresponding sides subtend.

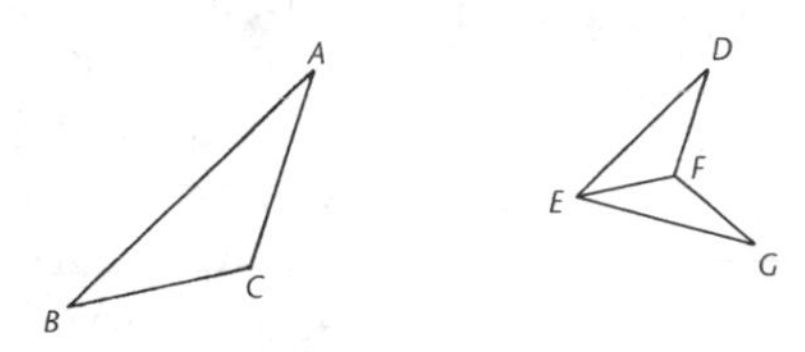

Proposition 6

If two triangles have one angle equal to one angle and the sides about the equal angles proportional, the triangles will be equiangular and will have those angles equal which the corresponding sides subtend.

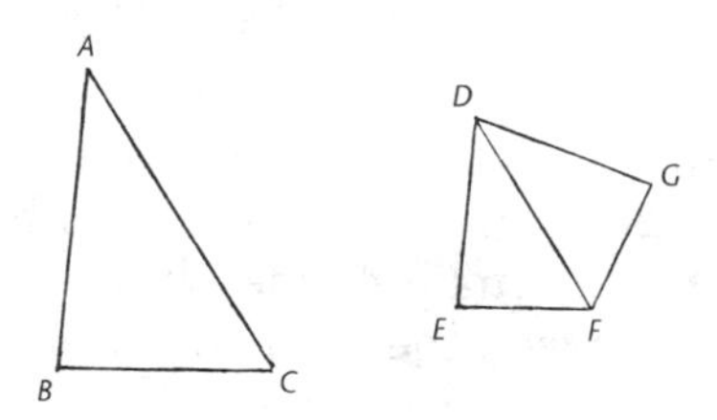

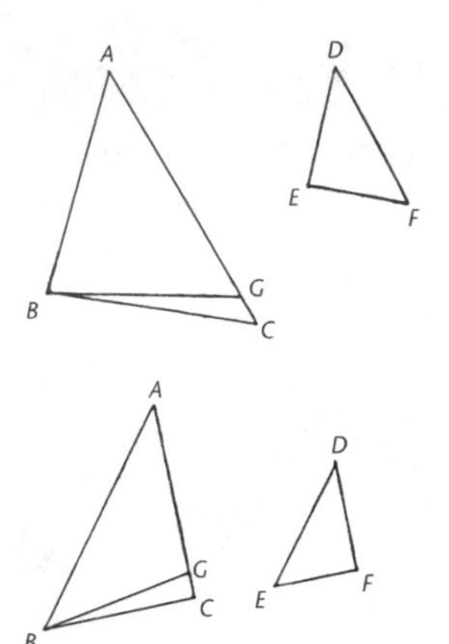

Proposition 7

If two triangles have one angle equal to one angle, the sides about other angles proportional, and the remaining angles either both less or both not less than a right angle, the triangles will be equiangular and will have those angles equal, the sides about which are proportional.

Proposition 8

If in a right-angled triangle a perpendicular be drawn from the right angle to the base, the triangles adjoining the perpendicular are similar both to the whole and to one another.

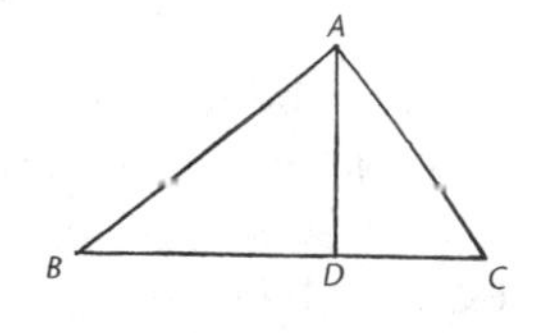

PORISM. From this it is clear that, if in a right-angled triangle a perpendicular be drawn from the right angle to the base, the straight line so drawn is a mean proportional between the segments of the base.

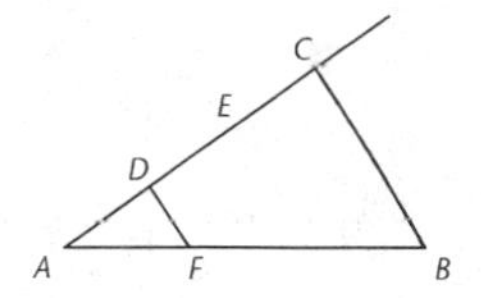

Proposition 9

From a given straight line to cut off a prescribed part.

Proposition 10

To cut a given uncut straight line similarly to a given cut straight line.

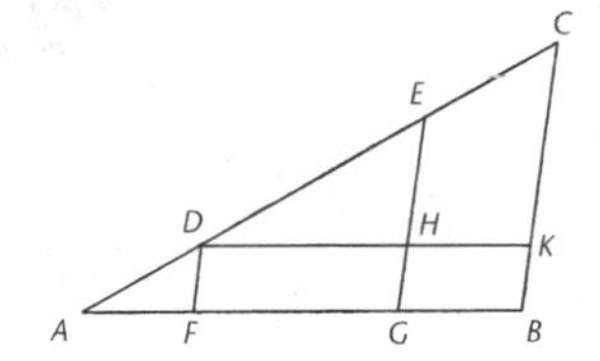

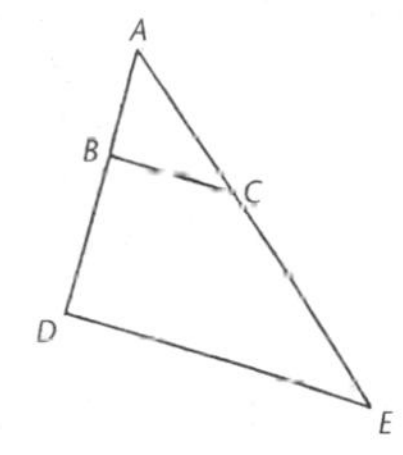

Proposition 11

To two given straight lines to find a third proportional.

Proposition 12

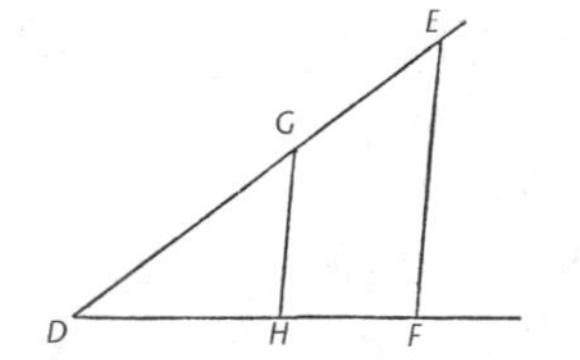

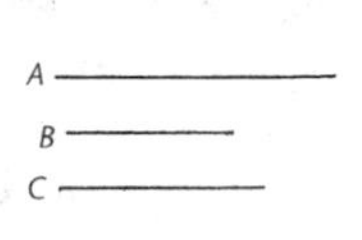

To three given straight lines to find a fourth proportional.

Proposition 13

To two given straight lines to find a mean proportional.

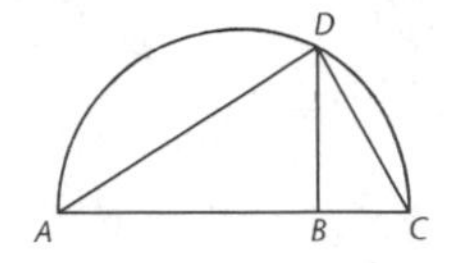

Proposition 14

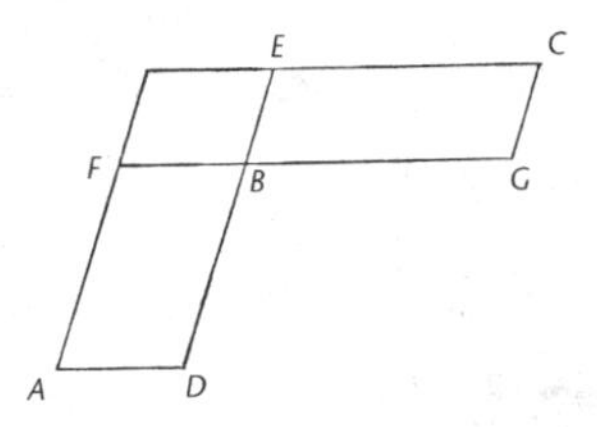

In equal and equiangular parallelograms the sides about the equal angles are reciprocally proportional; and equiangular parallelograms in which the sides about the equal angles are reciprocally proportional are equal.

Proposition 15

In equal triangles which have one angle equal to one angle the sides about the equal angles are reciprocally proportional; and those triangles which have one angle equal to one angle, and in which the sides about the equal angles are reciprocally proportional, are equal.

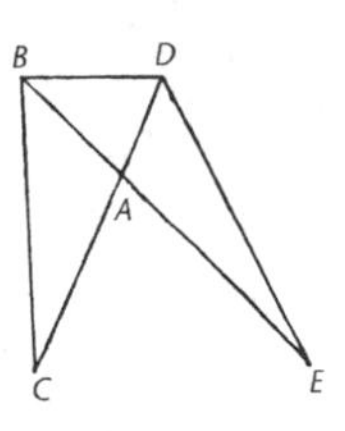

Proposition 16

If four straight lines be proportional, the rectangle contained by the extremes is equal to the rectangle contained by the means; and, if the rectangle contained by the extremes be equal to the rectangle contained by the means, the four straight lines will be proportional.

Proposition 17

If three straight lines be proportional, the rectangle contained by the extremes is equal to the square on the mean; and, if the rectangle contained by the extremes be equal to the square on the mean, the three straight lines will be proportional.

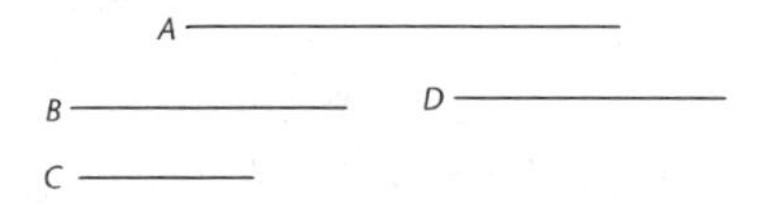

Proposition 18

On a given straight line to describe a rectilineal figure similar and similarly situated to a given rectilineal figure.

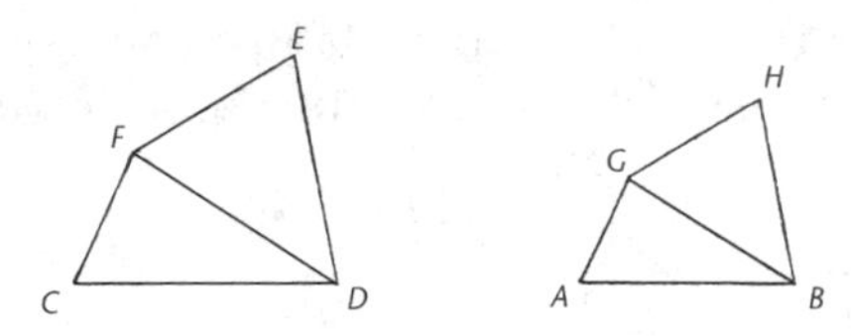

Proposition 19

Similar triangles are to one another in the duplicate ratio of the corresponding sides.

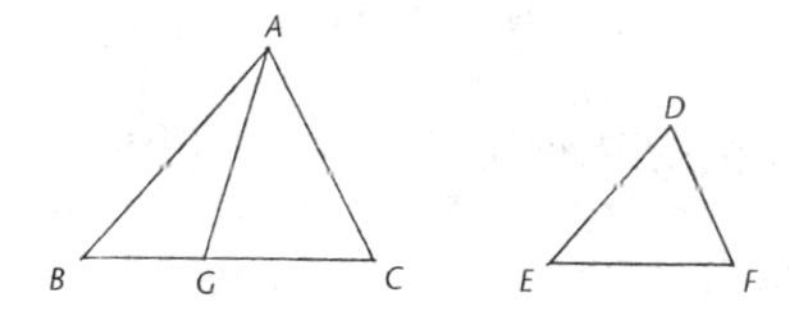

PORISM. From this it is manifest that, if three straight lines be proportional, then, as the first is to the third, so is the figure described on the first to that which is similar and similarly described on the second.

Proposition 20

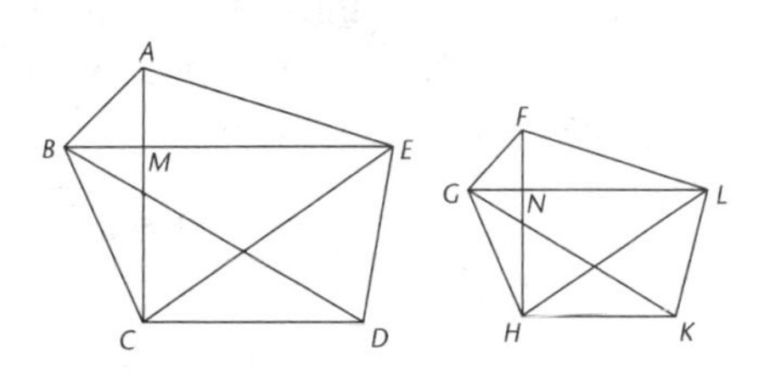

Similar polygons are divided into similar triangles, and into triangles equal in multitude and in the same ratio as the wholes, and the polygon has to the polygon a ratio duplicate of that which the corresponding side has to the corresponding side.

PORISM. Similarly also it can be proved in the case of quadrilaterals that they are in the duplicate ratio of the corresponding sides. And it was also proved in the case of triangles; therefore also, generally, similar rectilineal figures are to one another in the duplicate ratio of the corresponding sides.

Proposition 21

Figures which are similar to the same rectilineal figure are also similar to one another.

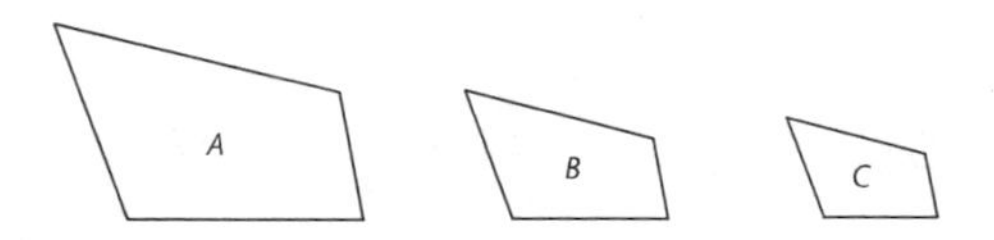

Proposition 22

If four straight lines be proportional, the rectilineal figures similar and similarly described upon them will also be proportional; and if the rectilineal figures similar and similarly described upon them be proportional, the straight lines will themselves also be proportional.

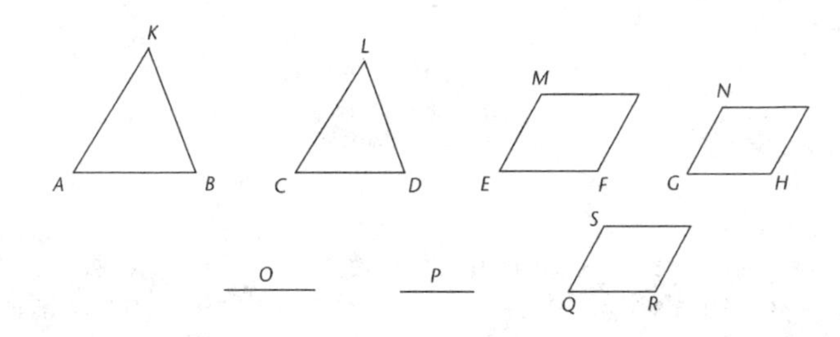

Proposition 23

Equiangular parallelograms have to one another the ratio compounded of the ratios of their sides.

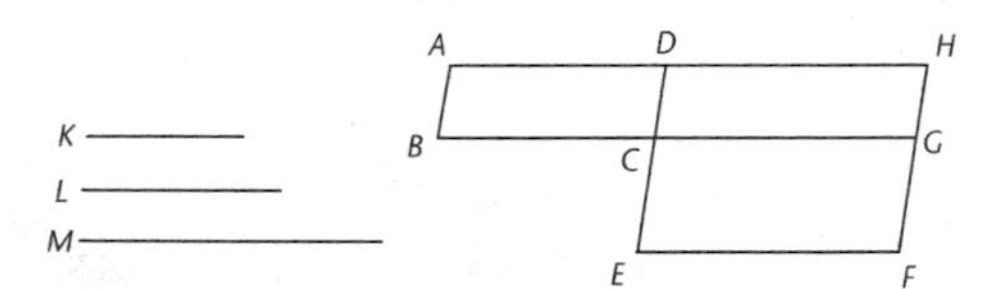

Proposition 24

In any parallelogram the parallelograms about the diameter are similar both to the whole and to one another.

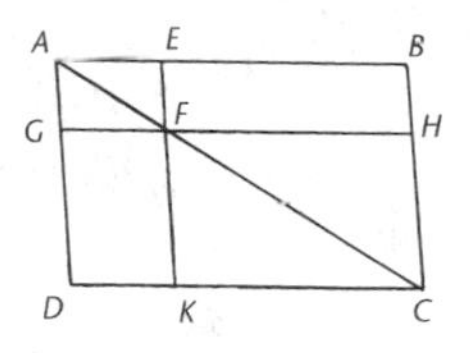

Proposition 25

To construct one and the same figure similar to a given rectilineal figure and equal to another given rectilineal figure.

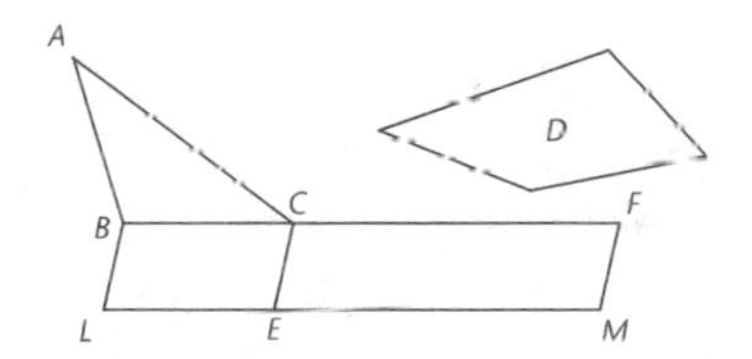

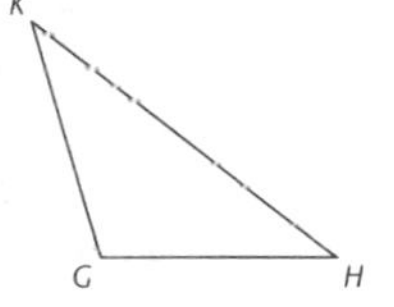

Proposition 26

If from a parallelogram there be taken away a parallelogram similar and similarly situated to the whole and having a common angle with it, it is about the same diameter with the whole.

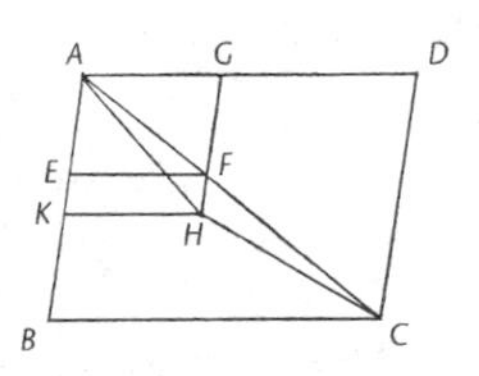

Proposition 27

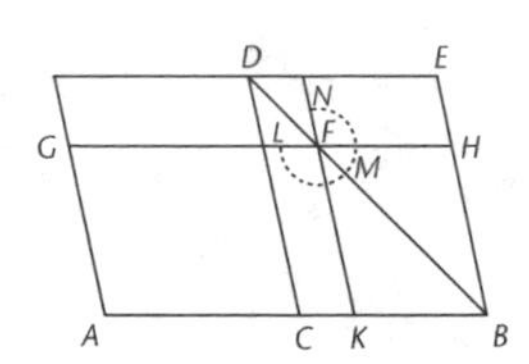

Of all the parallelograms applied to the same straight line and deficient by parallelogrammic figures similar and similarly situated to that described on the half of the straight line, that parallelogram is greatest which is applied to the half of the straight line and is similar to the defect.

Proposition 28

To a given straight line to apply a parallelogram equal to a given rectilineal figure and deficient by a parallelogrammic figure similar to a given one: thus the given rectilineal figure must not be greater than the parallelogram described on the half of the straight line and similar to the defect.

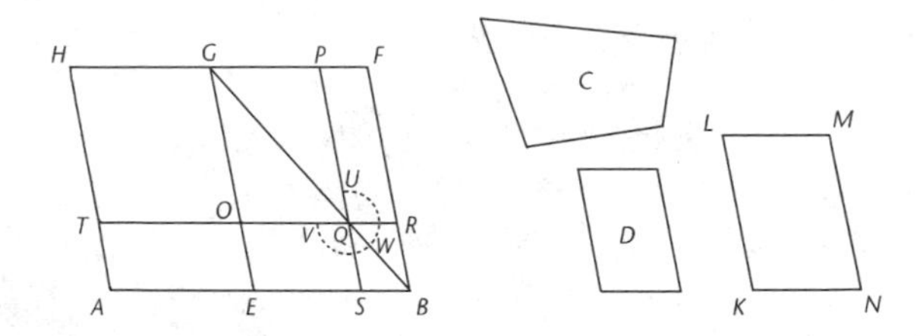

Proposition 29

To a given straight line to apply a parallelogram equal to a given rectilineal figure and exceeding by a parallelogrammic figure similar to a given one.

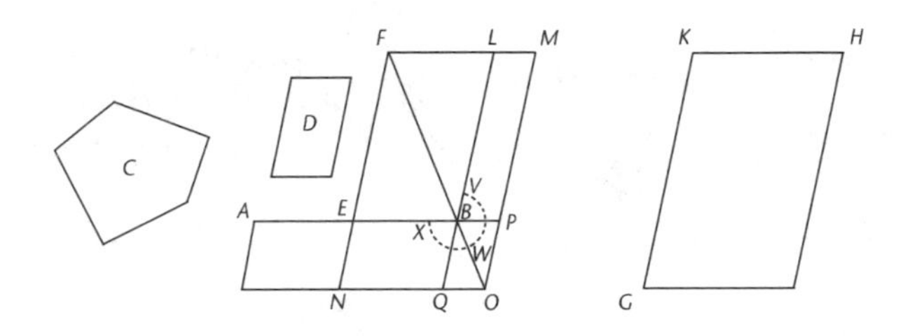

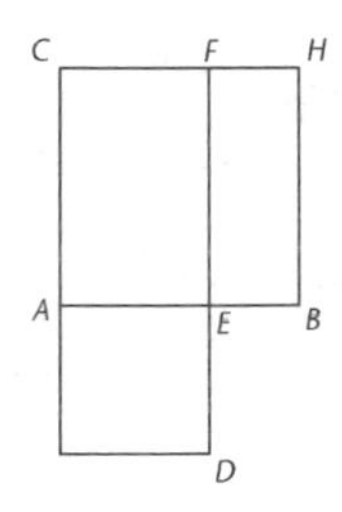

Proposition 30

To cut a given finite straight line in extreme and mean ratio.

Proposition 31

In right-angled triangles the figure on the side subtending the right angle is equal to the similar and similarly described figures on the sides containing the right angle.

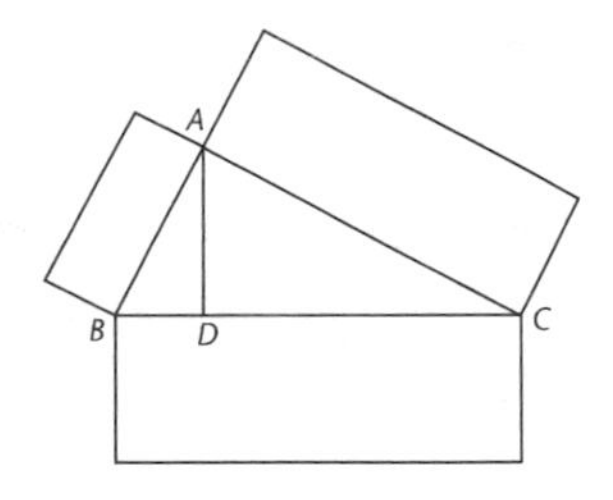

Proposition 32

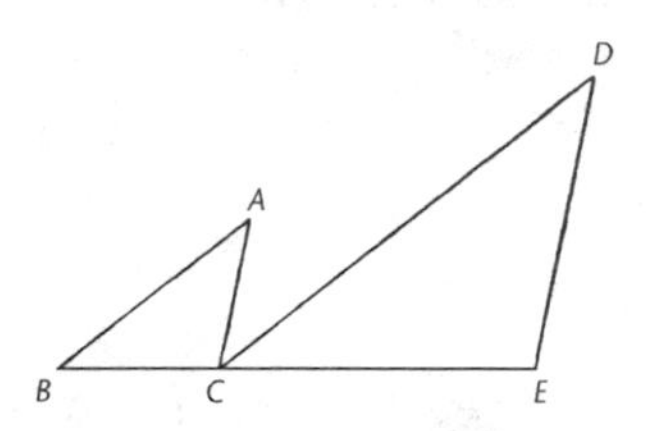

If two triangles having two sides proportional to two sides be placed together at one angle so that their corresponding sides are also parallel, the remaining sides of the triangles will be in a straight line.

Proposition 33

In equal circles angles have the same ratio as the circumferences on which they stand, whether they stand at the centres or at the circumferences.

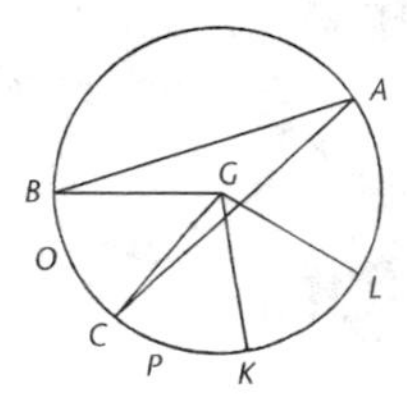

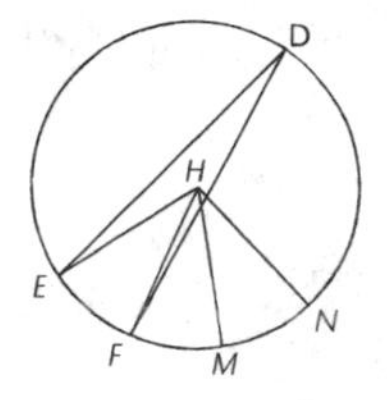

Book VII

Definitions

1. An *unit* is that by virtue of which each of the things that exist is called one.
2. A *number* is a multitude composed of units.
3. A number is *a part* of a number, the less of the greater, when it measures the greater;
4. but *parts* when it does not measure it.
5. The greater number is a *multiple* of the less when it is measured by the less.
6. An *even number* is that which is divisible into two equal parts.
7. An *odd number* is that which is not divisible into two equal parts, or that which differs by an unit from an even number.
8. An *even-times even number* is that which is measured by an even number according to an even number.
9. An *even-times odd number* is that which is measured by an even number according to an odd number.
10. An *odd-times odd number* is that which is measured by an odd number according to an odd number.
11. A *prime number* is that which is measured by an unit alone.
12. Numbers *prime to one another* are those which are measured by an unit alone as a common measure.

13. A *composite number* is that which is measured by some number.
14. Numbers *composite to one another* are those which are measured by some number as a common measure.
15. A number is said to *multiply* a number when that which is multiplied is added to itself as many times as there are units in the other, and thus some number is produced.
16. And, when two numbers having multiplied one another make some number, the number so produced is called *plane,* and its *sides* are the numbers which have multiplied one another.
17. And, when three numbers having multiplied one another make some number, the number so produced is *solid,* and its sides are the numbers which have multiplied one another.
18. A *square number* is equal multiplied by equal, or a number which is contained by two equal numbers.
19. And a *cube* is equal multiplied by equal and again by equal, or a number which is contained by three equal numbers.
20. Numbers are *proportional* when the first is the same multiple, or the same part, or the same parts, of the second that the third is of the fourth.
21. *Similar plane* and *solid* numbers are those which have their sides proportional.
22. A *perfect number* is that which is equal to its own parts.

Proposition 1

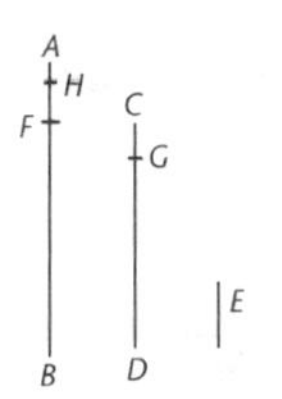

Two unequal numbers being set out, and the less being continually subtracted in turn from the greater, if the number which is left never measures the one before it until an unit is left, the original numbers will be prime to one another.

Proposition 2

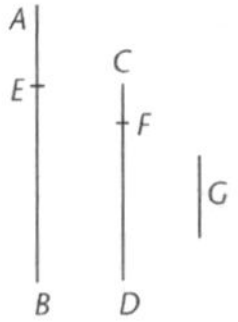

Given two numbers not prime to one another, to find their greatest common measure.

PORISM. From this it is manifest that, if a number measure two numbers, it will also measure their greatest common measure.

Proposition 3

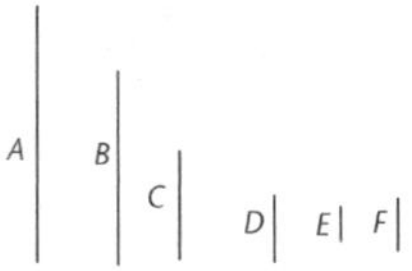

Given three numbers not prime to one another, to find their greatest common measure.

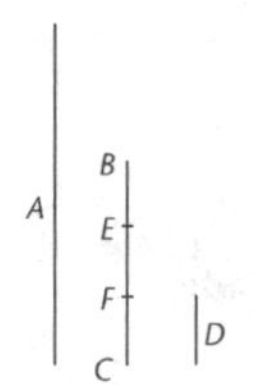

Proposition 4

Any number is either a part or parts of any number, the less of the greater.

Proposition 5

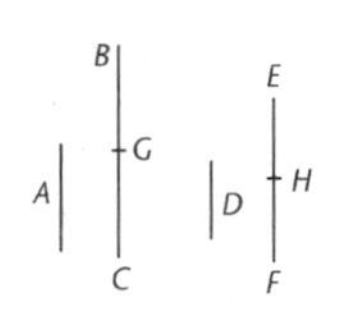

If a number be a part of a number, and another be the same part of another, the sum will also be the same part of the sum that the one is of the one.

Proposition 6

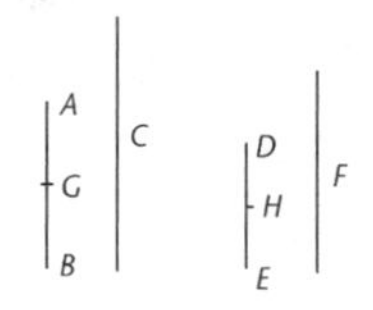

If a number be parts of a number, and another be the same parts of another, the sum will also be the same parts of the sum that the one is of the one.

Proposition 7

If a number be that part of a number, which a number subtracted is of a number subtracted, the remainder will also be the same part of the remainder that the whole is of the whole.

Proposition 8

If a number be the same parts of a number that a number subtracted is of a number subtracted, the remainder will also be the same parts of the remainder that the whole is of the whole.

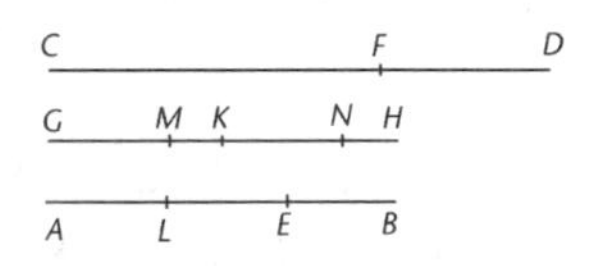

Proposition 9

If a number be a part of a number, and another be the same part of another, alternately also, whatever part or parts the first is of the third, the same part, or the same parts, will the second also be of the fourth.

Proposition 10

If a number be parts of a number, and another be the same parts of another, alternately also, whatever parts or part the first is of the third, the same parts or the same part will the second also be of the fourth.

Proposition 11

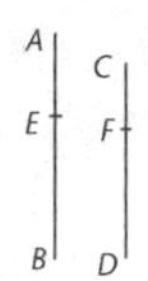

If, as whole is to whole, so is a number subtracted to a number subtracted, the remainder will also be to the remainder as whole to whole.

Proposition 12

If there be as many numbers as we please in proportion, then, as one of the antecedents is to one of the consequents, so are all the antecedents to all the consequents.

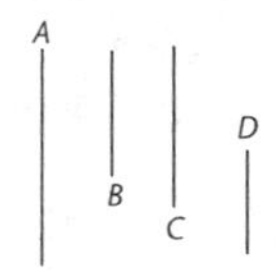

Proposition 13

If four numbers be proportional, they will also be proportional alternately.

Proposition 14

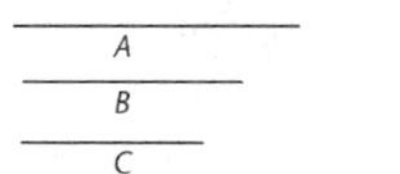

If there be as many numbers as we please, and others equal to them in multitude, which taken two and two are in the same ratio, they will also be in the same ratio *ex aequali*.

Proposition 15

If an unit measure any number, and another number measure any other number the same number of times, alternately also, the unit will measure the third number the same number of times that the second measures the fourth.

Proposition 16

If two numbers by multiplying one another make certain numbers, the numbers so produced will be equal to one another.

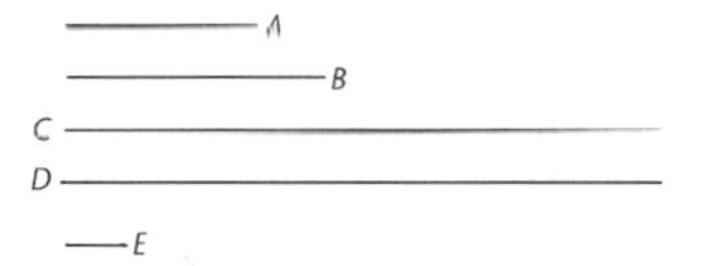

Proposition 17

If a number by multiplying two numbers make certain numbers, the numbers so produced will have the same ratio as the numbers multiplied.

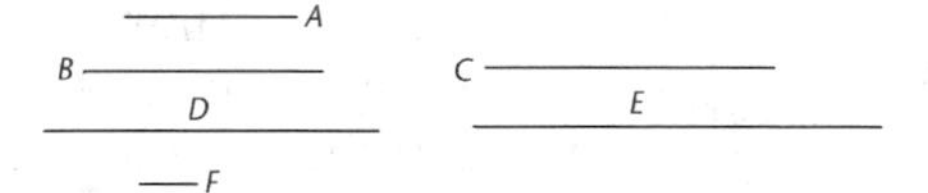

Proposition 18

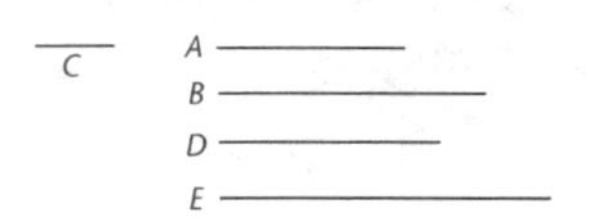

If two numbers by multiplying any number make certain numbers, the numbers so produced will have the same ratio as the multipliers.

Proposition 19

If four numbers be proportional, the number produced from the first and fourth will be equal to the number produced from the second and third; and, if the number produced from the first and fourth be equal to that produced from the second and third, the four numbers will be proportional.

Proposition 20

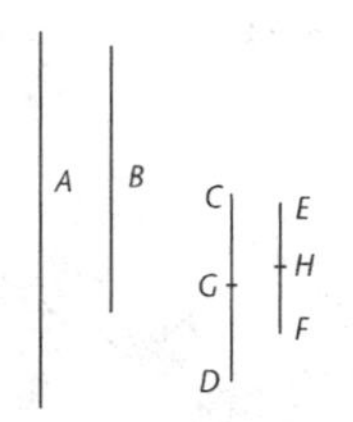

The least numbers of those which have the same ratio with them measure those which have the same ratio the same number of times, the greater the greater and the less the less.

Proposition 21

Numbers prime to one another are the least of those which have the same ratio with them.

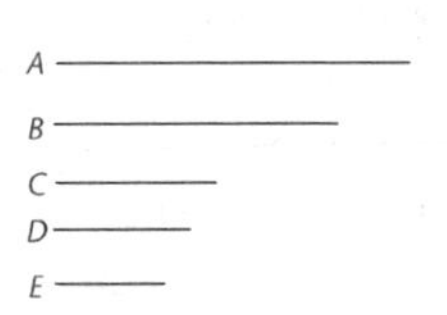

Proposition 22

The least numbers of those which have the same ratio with them are prime to one another.

Proposition 23

If two numbers be prime to one another, the number which measures the one of them will be prime to the remaining number.

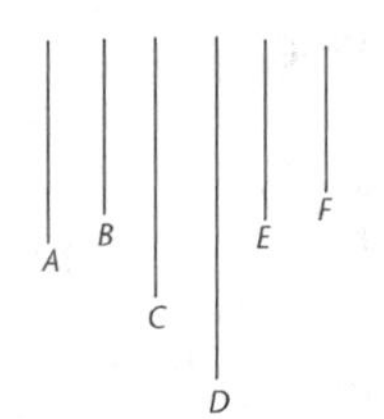

Proposition 24

If two numbers be prime to any number, their product also will be prime to the same.

Proposition 25

If two numbers be prime to one another, the product of one of them into itself will be prime to the remaining one.

Proposition 26

If two numbers be prime to two numbers, both to each, their products also will be prime to one another.

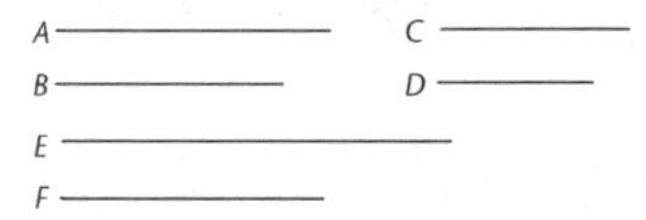

Proposition 27

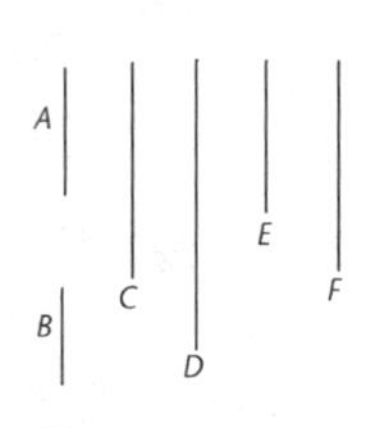

If two numbers be prime to one another, and each by multiplying itself make a certain number, the products will be prime to one another; and, if the original numbers by multiplying the products make certain numbers, the latter will also be prime to one another ⟨and this is always the case with the extremes⟩.

Proposition 28

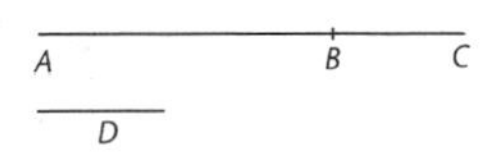

If two numbers be prime to one another, the sum will also be prime to each of them; and, if the sum of two numbers be prime to any one of them, the original numbers will also be prime to one another.

Proposition 29

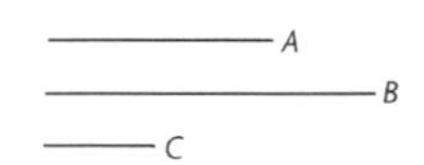

Any prime number is prime to any number which it does not measure.

Proposition 30

If two numbers by multiplying one another make some number, and any prime number measure the product, it will also measure one of the original numbers.

Proposition 31

Any composite number is measured by some prime number.

Proposition 32

Any number either is prime or is measured by some prime number.

Proposition 33

Given as many numbers as we please, to find the least of those which have the same ratio with them.

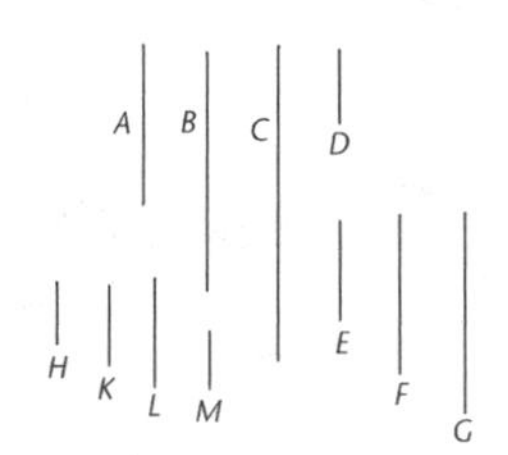

Proposition 34

Given two numbers, to find the least number which they measure.

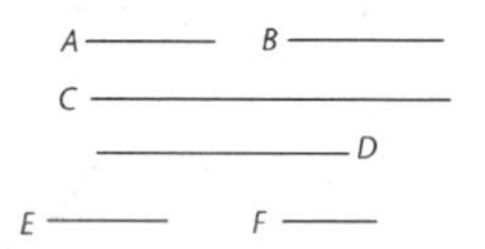

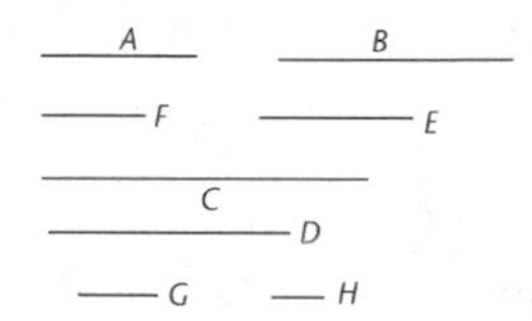

Proposition 35

If two numbers measure any number, the least number measured by them will also measure the same.

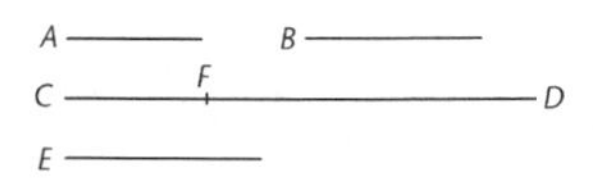

Proposition 36

Given three numbers, to find the least number which they measure.

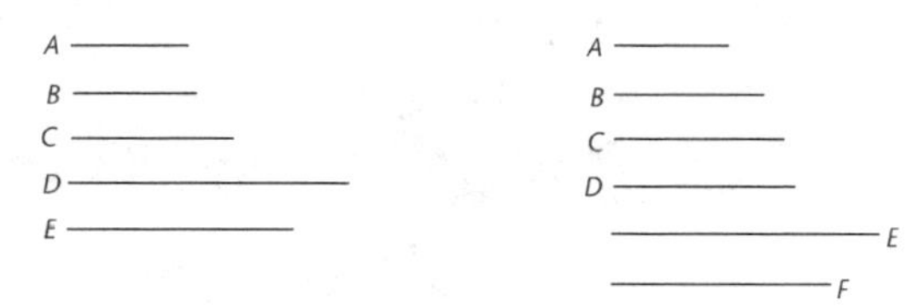

Proposition 37

If a number be measured by any number, the number which is measured will have a part called by the same name as the measuring number.

Proposition 38

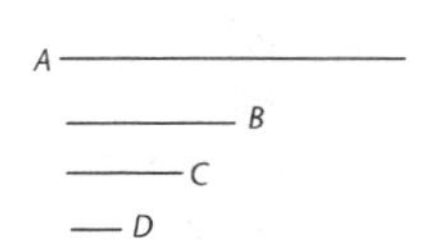

If a number have any part whatever, it will be measured by a number called by the same name as the part.

Proposition 39

To find the number which is the least that will have given parts.

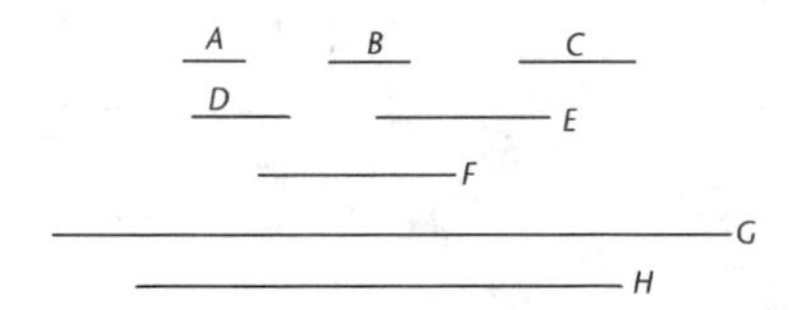

Book VIII

Proposition 1

If there be as many numbers as we please in continued proportion, and the extremes of them be prime to one another, the numbers are the least of those which have the same ratio with them.

Proposition 2

To find numbers in continued proportion, as many as may be prescribed, and the least that are in a given ratio.

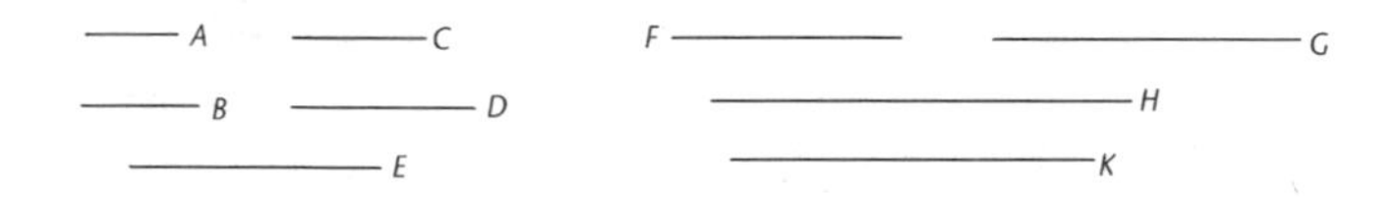

PORISM. From this it is manifest that, if three numbers in continued proportion be the least of those which have the same ratio with them, the extremes of them are squares, and, if four numbers, cubes.

Proposition 3

If as many numbers as we please in continued proportion be the least of those which have the same ratio with them, the extremes of them are prime to one another.

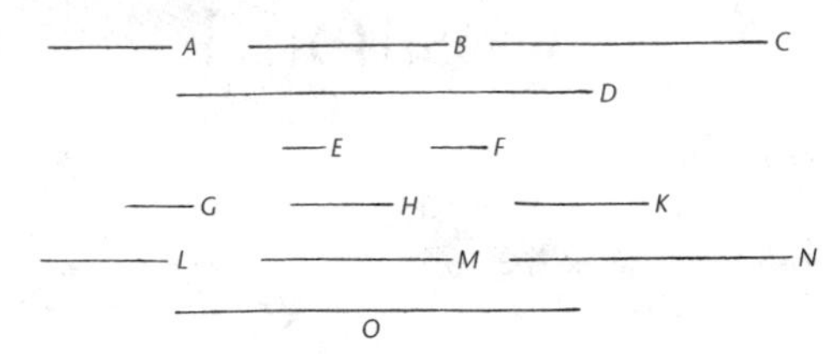

Proposition 4

Given as many ratios as we please in least numbers, to find numbers in continued proportion which are the least in the given ratios.

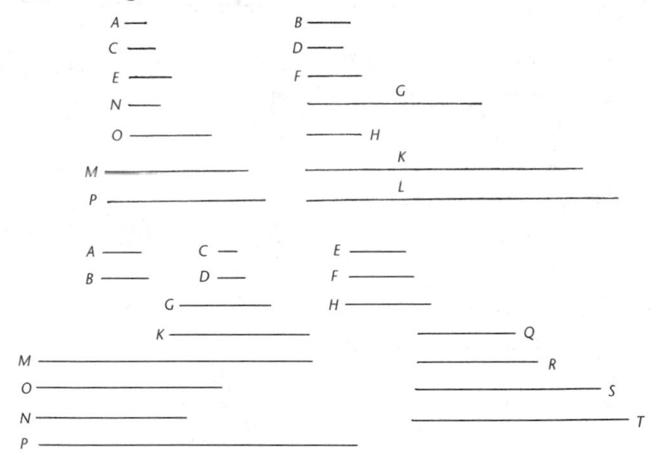

Proposition 5

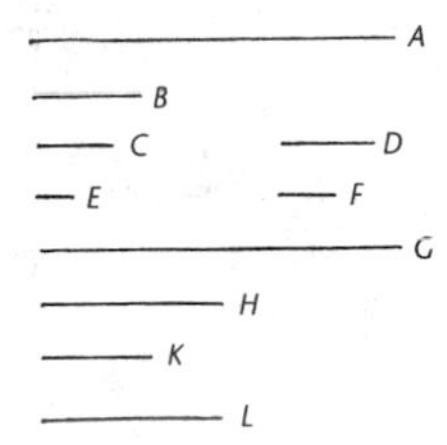

Plane numbers have to one another the ratio compounded of the ratios of their sides.

Proposition 6

If there be as many numbers as we please in continued proportion, and the first do not measure the second, neither will any other measure any other.

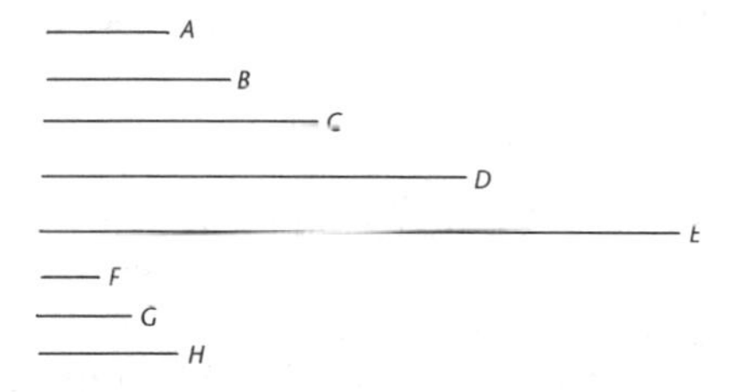

Proposition 7

If there be as many numbers as we please in continued proportion, and the first measure the last, it will measure the second also.

A
B
C
D

Proposition 8

If between two numbers there fall numbers in continued proportion with them, then, however many numbers fall between them in continued proportion, so many will also fall in continued proportion between the numbers which have the same ratio with the original numbers.

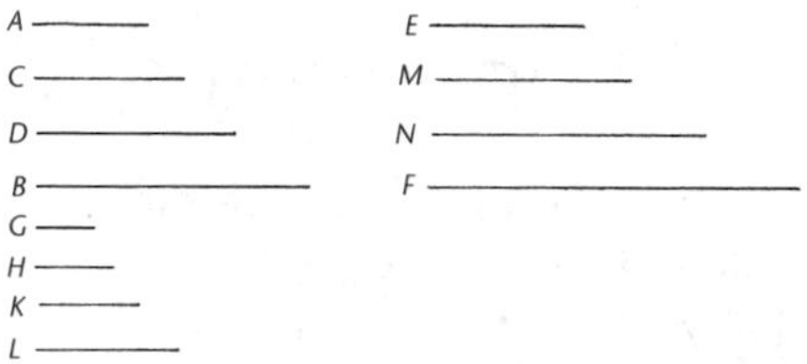

A, C, D, B, G, H, K, L, E, M, N, F

Proposition 9

If two numbers be prime to one another, and numbers fall between them in continued proportion, then, however many numbers fall between them in continued proportion, so many will also fall between each of them and an unit in continued proportion.

A, C, D, B, E, F, G, H, K, L, M, N, O, P

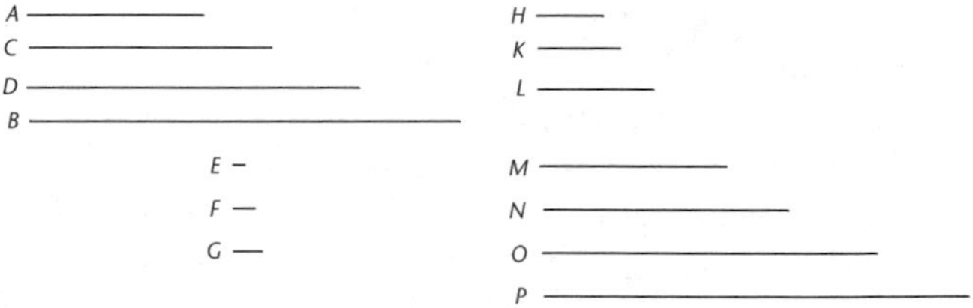

Proposition 10

If numbers fall between each of two numbers and an unit in continued proportion, however many numbers fall between each of them and an unit in continued proportion, so many also will fall between the numbers themselves in continued proportion.

Proposition 11

Between two square numbers there is one mean proportional number, and the square has to the square the ratio duplicate of that which the side has to the side.

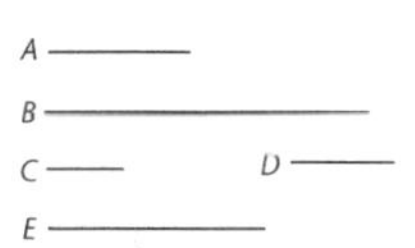

Proposition 12

Between two cube numbers there are two mean proportional numbers, and the cube has to the cube the ratio triplicate of that which the side has to the side.

Proposition 13

If there be as many numbers as we please in continued proportion, and each by multiplying itself make some number, the products will be proportional; and, if the original numbers by multiplying the products make certain numbers, the latter will also be proportional.

A G
B H
C K
D
E M
F N
L P
O Q

Proposition 14

If a square measure a square, the side will also measure the side; and, if the side measure the side, the square will also measure the square.

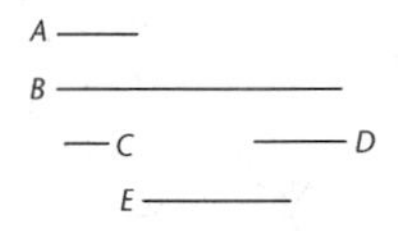

Proposition 15

If a cube number measure a cube number, the side will also measure the side; and, if the side measure the side, the cube will also measure the cube.

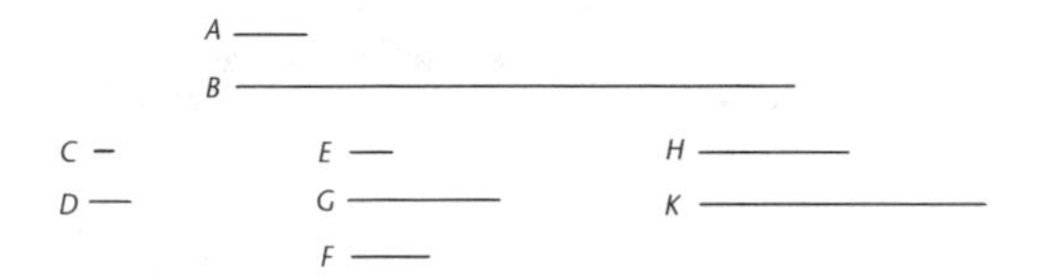

Proposition 16

If a square number do not measure a square number, neither will the side measure the side; and, if the side do not measure the side, neither will the square measure the square.

Proposition 17

If a cube number do not measure a cube number, neither will the side measure the side; and, if the side do not measure the side, neither will the cube measure the cube.

Proposition 18

Between two similar plane numbers there is one mean proportional number; and the plane number has to the plane number the ratio duplicate of that which the corresponding side has to the corresponding side.

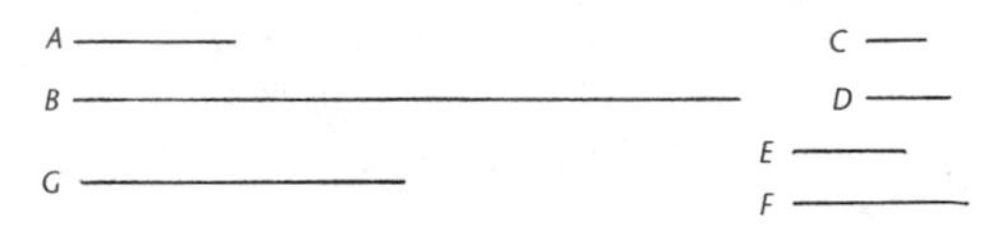

Proposition 19

Between two similar solid numbers there fall two mean proportional numbers; and the solid number has to the similar solid number the ratio triplicate of that which the corresponding side has to the corresponding side.

A
B
C F N
D G O
E H
K
L
M

Proposition 20

If one mean proportional number fall between two numbers, the numbers will be similar plane numbers.

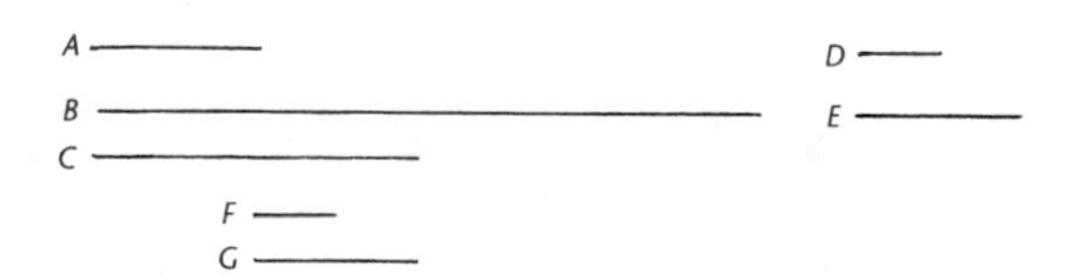

Proposition 21

If two mean proportional numbers fall between two numbers, the numbers are similar solid numbers.

Proposition 22

If three numbers be in continued proportion, and the first be square, the third will also be square.

Proposition 23

If four numbers be in continued proportion, and the first be cube, the fourth will also be cube.

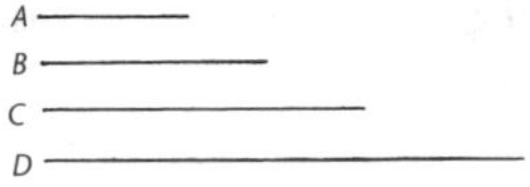

A
B
C
D

Proposition 24

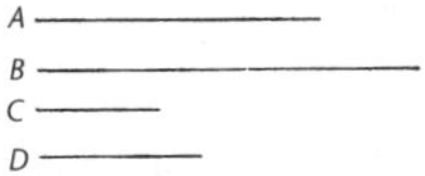

A
B
C
D

If two numbers have to one another the ratio which a square number has to a square number, and the first be square, the second will also be square.

Proposition 25

If two numbers have to one another the ratio which a cube number has to a cube number, and the first be cube, the second will also be cube.

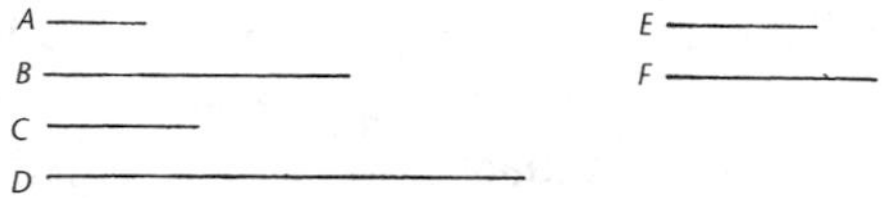

A
B
C
D

E
F

Proposition 26

Similar plane numbers have to one another the ratio which a square number has to a square number.

Proposition 27

Similar solid numbers have to one another the ratio which a cube number has to a cube number.

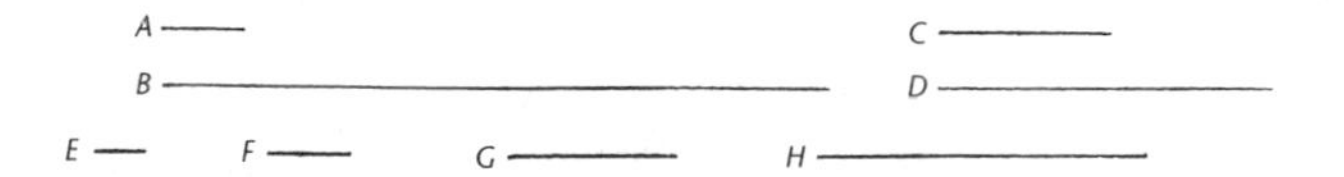

Book IX

Proposition 1

If two similar plane numbers by multiplying one another make some number, the product will be square.

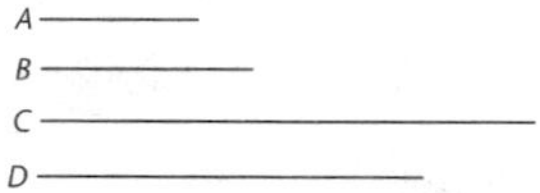

A
B
C
D

Proposition 2

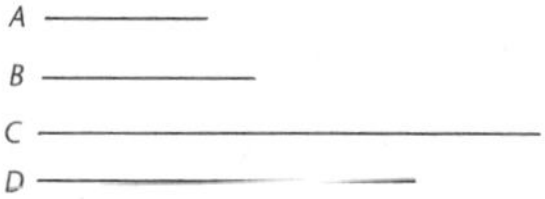

A
B
C
D

If two numbers by multiplying one another make a square number, they are similar plane numbers.

Proposition 3

If a cube number by multiplying itself make some number, the product will be cube.

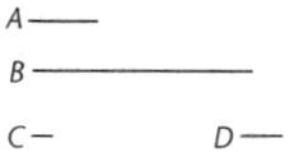

A
B
C D

Proposition 4

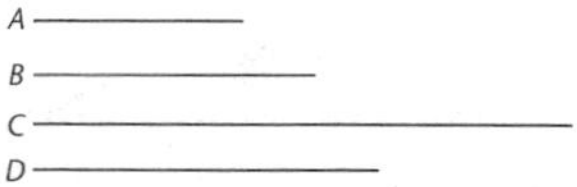

A
B
C
D

If a cube number by multiplying a cube number make some number, the product will be cube.

Proposition 5

If a cube number by multiplying any number make a cube number, the multiplied number will also be cube.

Proposition 6

If a number by multiplying itself make a cube number, it will itself also be cube.

Proposition 7

If a composite number by multiplying any number make some number, the product will be solid.

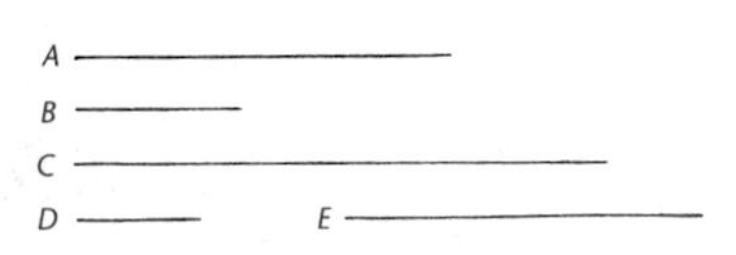

Proposition 8

If as many numbers as we please beginning from an unit be in continued proportion, the third from the unit will be square, as will also those which successively leave out one; the fourth will be cube, as will also all those which leave out two; and the seventh will be at once cube and square, as will also those which leave out five.

Proposition 9

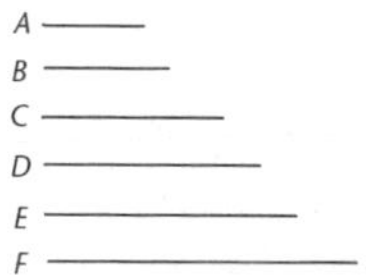

A
B
C
D
E
F

If as many numbers as we please beginning from an unit be in continued proportion, and the number after the unit be square, all the rest will also be square. And, if the number after the unit be cube, all the rest will also be cube.

Proposition 10

If as many numbers as we please beginning from an unit be in continued proportion, and the number after the unit be not square, neither will any other be square except the third from the unit and all those which leave out one. And, if the number after the unit be not cube, neither will any other be cube except the fourth from the unit and all those which leave out two.

A
B
C
D
E
F

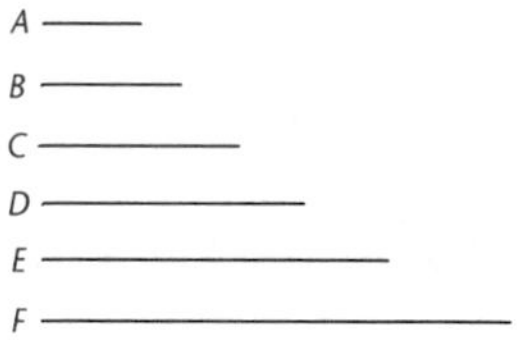

Proposition 11

If as many numbers as we please beginning from an unit be in continued proportion, the less measures the greater according to some one of the numbers which have place among the proportional numbers.

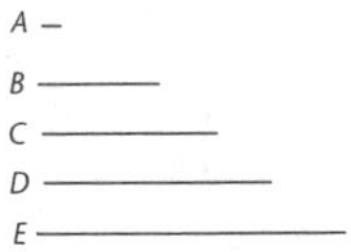

A
B
C
D
E

PORISM. And it is manifest that, whatever place the measuring number has, reckoned from the unit, the same place also has the number according to which it measures, reckoned from the number measured, in the direction of the number before it.

Proposition 12

If as many numbers as we please beginning from an unit be in continued proportion, by however many prime numbers the last is measured, the next to the unit will also be measured by the same.

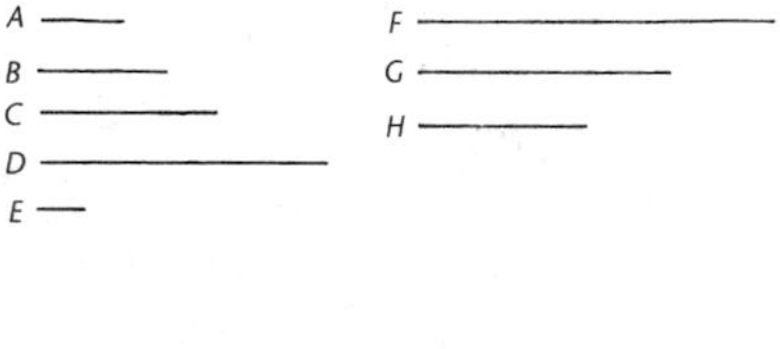

A
B
C
D
E
F
G
H

Proposition 13

If as many numbers as we please beginning from an unit be in continued proportion, and the number after the unit be prime, the greatest will not be measured by any except those which have a place among the proportional numbers.

Proposition 14

If a number be the least that is measured by prime numbers, it will not be measured by any other prime number except those originally measuring it.

Proposition 15

If three numbers in continued proportion be the least of those which have the same ratio with them, any two whatever added together will be prime to the remaining number.

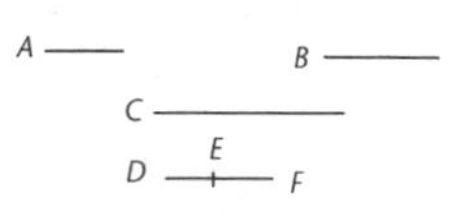

Proposition 16

If two numbers be prime to one another, the second will not be to any other number as the first is to the second.

A
B
C

Proposition 17

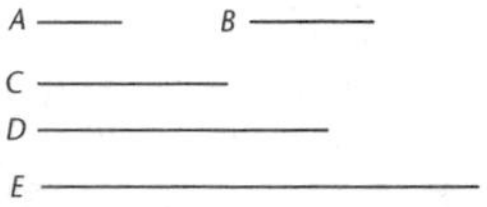

A B
C
D
E

If there be as many numbers as we please in continued proportion, and the extremes of them be prime to one another, the last will not be to any other number as the first to the second.

Proposition 18

Given two numbers, to investigate whether it is possible to find a third proportional to them.

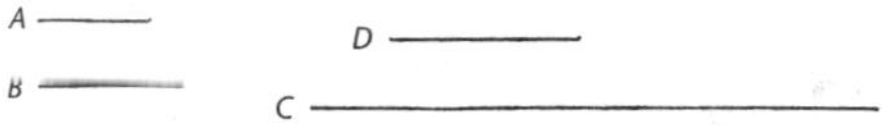

A
D
B
C

Proposition 19

Given three numbers, to investigate when it is possible to find a fourth proportional to them.

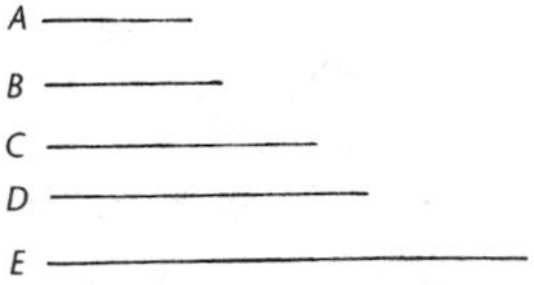

A
B
C
D
E

Proposition 20

Prime numbers are more than any assigned multitude of prime numbers.

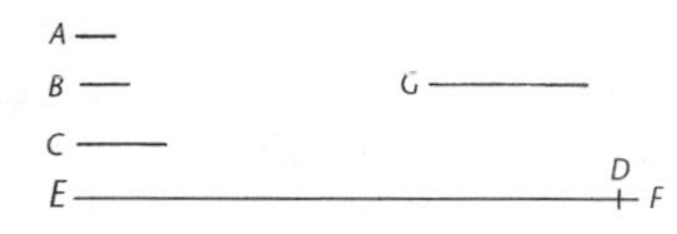

Proposition 21

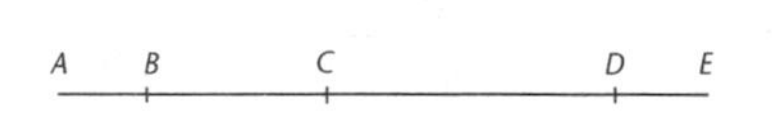

If as many even numbers as we please be added together, the whole is even.

Proposition 22

If as many odd numbers as we please be added together, and their multitude be even, the whole will be even.

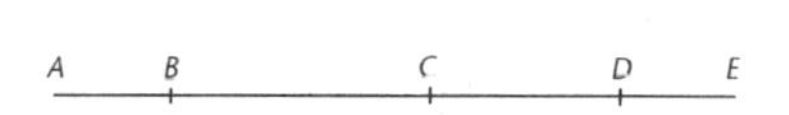

Proposition 23

If as many odd numbers as we please be added together, and their multitude be odd, the whole will also be odd.

Proposition 24

If from an even number an even number be subtracted, the remainder will be even.

Proposition 25

If from an even number an odd number be subtracted, the remainder will be odd.

Proposition 26

If from an odd number an odd number be subtracted, the remainder will be even.

Proposition 27

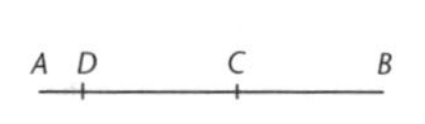

If from an odd number an even number be subtracted, the remainder will be odd.

Proposition 28

If an odd number by multiplying an even number make some number, the product will be even.

A
B
C

Proposition 29

If an odd number by multiplying an odd number make some number, the product will be odd.

A
B
C

Proposition 30

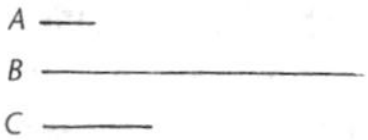

A
B
C

If an odd number measure an even number, it will also measure the half of it.

Proposition 31

If an odd number be prime to any number, it will also be prime to the double of it.

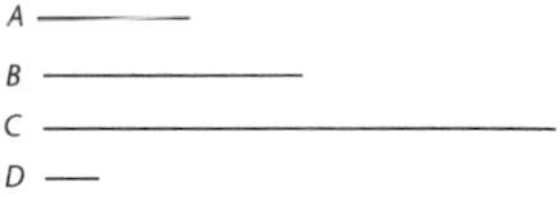

A
B
C
D

Proposition 32

A
B
C
D

Each of the numbers which are continually doubled beginning from a dyad is even-times even only.

Proposition 33

If a number have its half odd, it is even-times odd only.

A

Proposition 34

If a number neither be one of those which are continually doubled from a dyad, nor have its half odd, it is both even-times even and even-times odd.

A

Proposition 35

If as many numbers as we please be in continued proportion, and there be subtracted, from the second and the last, numbers equal to the first, then, as the excess of the second is to the first, so will the excess of the last be to all those before it.

A
B G C
D
E L K H F

Proposition 36

If as many numbers as we please beginning from an unit be set out continuously in double proportion, until the sum of all becomes prime, and if the sum multiplied into the last make some number, the product will be perfect.

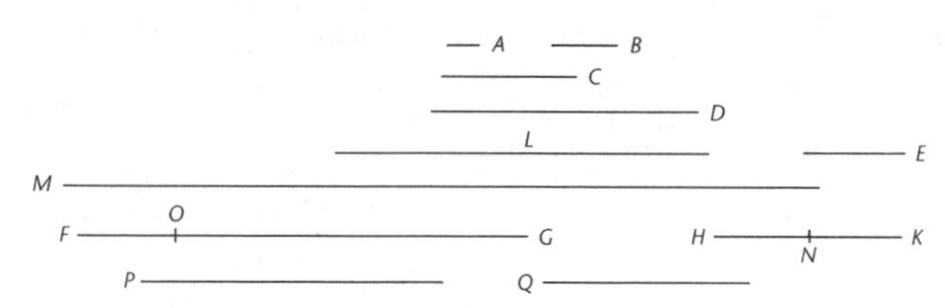

Book X

Definitions I[1]

1. Those magnitudes are said to be *commensurable* which are measured by the same measure, and those *incommensurable* which cannot have any common measure.

2. Straight lines are *commensurable in square* when the squares on them are measured by the same area, and *incommensurable in square* when the squares on them cannot possibly have any area as a common measure.

3. With these hypotheses, it is proved that there exist straight lines infinite in multitude which are commensurable and incommensurable respectively, some in length only, and others in square also, with an assigned straight line. Let then the assigned straight line be called *rational,* and those straight lines which are commensurable with it, whether in length and in square or in square only, *rational,* but those which are incommensurable with it *irrational.*

4. And let the square on the assigned straight line be called *rational* and those areas which are commensurable with it *rational,* but those which are incommensurable with it *irrational,* and the straight lines

1. A second set of Definitions for Book X begins on page 132, a third on page 146. —Ed.

which produce them *irrational,* that is, in case the areas are squares, the sides themselves, but in case they are any other rectilineal figures, the straight lines on which are described squares equal to them.

Proposition 1

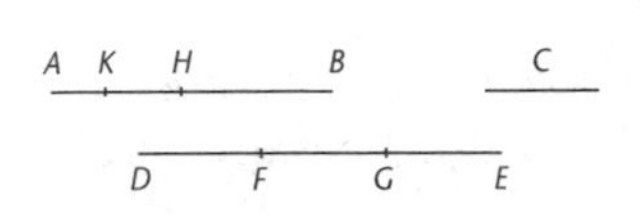

Two unequal magnitudes being set out, if from the greater there be subtracted a magnitude greater than its half, and from that which is left a magnitude greater than its half, and if this process be repeated continually, there will be left some magnitude which will be less than the lesser magnitude set out.

PORISM. And the theorem can be similarly proved even if the parts subtracted be halves.

Proposition 2

If, when the less of two unequal magnitudes is continually subtracted in turn from the greater, that which is left never measures the one before it, the magnitudes will be incommensurable.

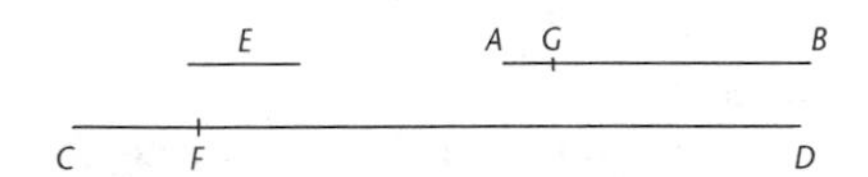

Proposition 3

Given two commensurable magnitudes, to find their greatest common measure.

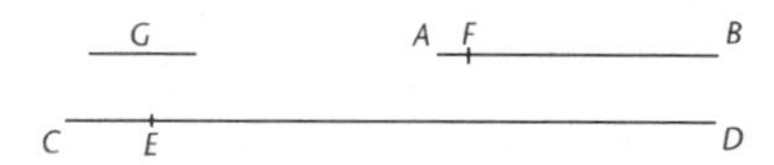

PORISM. From this it is manifest that, if a magnitude measure two magnitudes, it will also measure their greatest common measure.

Proposition 4

Given three commensurable magnitudes, to find their greatest common measure.

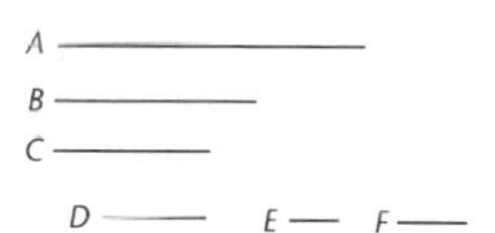

Proposition 5

Commensurable magnitudes have to one another the ratio which a number has to a number.

Proposition 6

If two magnitudes have to one another the ratio which a number has to a number, the magnitudes will be commensurable.

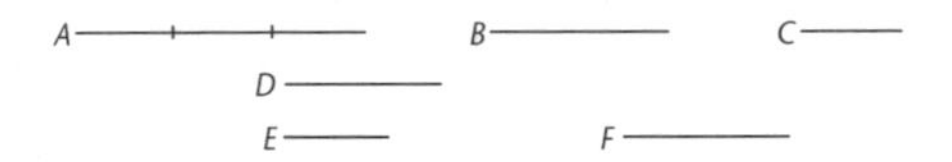

PORISM. From this it is manifest that, if there be two numbers, as *D, E,* and a straight line, as *A,* it is possible to make a straight line [*F*] such that the given straight line is to it as the number *D* is to the number *E*.

Proposition 7

Incommensurable magnitudes have not to one another the ratio which a number has to a number.

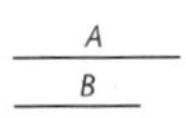

Proposition 8

If two magnitudes have not to one another the ratio which a number has to a number, the magnitudes will be incommensurable.

Proposition 9

The squares on straight lines commensurable in length have to one another the ratio which a square number has to a square number; and squares which have to one another the ratio which a square number has to a square number will also have their sides commensurable in length. But the squares on straight lines incommensurable in length have not to one another the ratio which a square number has to a square number; and squares which have not to one another the ratio which a square number has to a square number will not have their sides commensurable in length either.

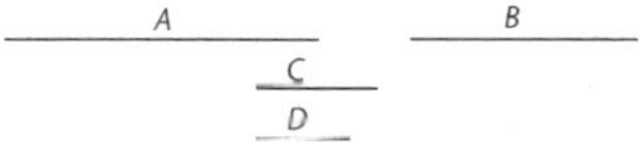

A B

C

D

PORISM. And it is manifest from what has been proved that straight lines commensurable in length are always commensurable in square also, but those commensurable in square are not always commensurable in length also.

⟨**LEMMA.** Numbers which are not similar plane numbers, that is, those which have not their sides proportional, have not to one another the ratio which a square number has to a square number.⟩

⟨Proposition 10

A
D
E
B
C

To find two straight lines incommensurable, the one in length only, and the other in square also, with an assigned straight line.⟩

Proposition 11

If four magnitudes be proportional, and the first be commensurable with the second, the third will also be commensurable with the fourth; and, if the first be incommensurable with the second, the third will also be incommensurable with the fourth.

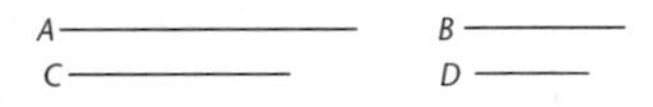

Proposition 12

Magnitudes commensurable with the same magnitude are commensurable with one another also.

A C B
D
E H
F K
G L

Proposition 13

If two magnitudes be commensurable, and the one of them be incommensurable with any magnitude, the remaining one will also be incommensurable with the same.

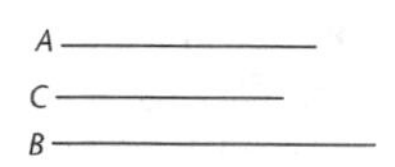

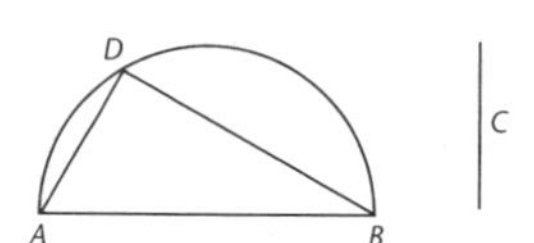

LEMMA. Given two unequal straight lines, to find by what square the square on the greater is greater than the square on the less.

Proposition 14

If four straight lines be proportional, and the square on the first be greater than the square on the second by the square on a straight line commensurable with the first, the square on the third will also be greater than the square on the fourth by the square on a straight line commensurable with the third.

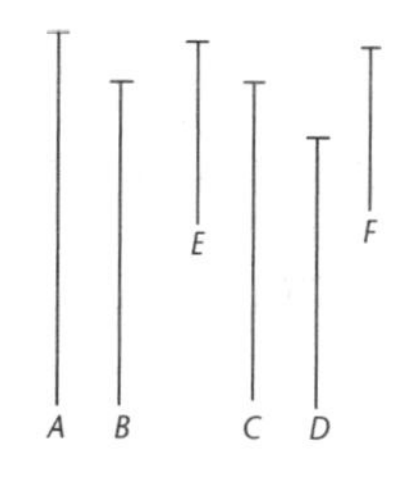

And, if the square on the first be greater than the square on the second by the square on a straight line incommensurable with the first, the square on the third will also be greater than the square on the fourth by the square on a straight line incommensurable with the third.

Proposition 15

If two commensurable magnitudes be added together, the whole will also be commensurable with each of them; and, if the whole be commensurable with one of them, the original magnitudes will also be commensurable.

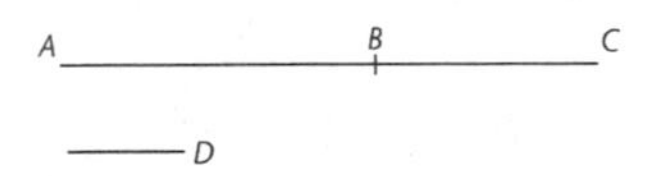

Proposition 16

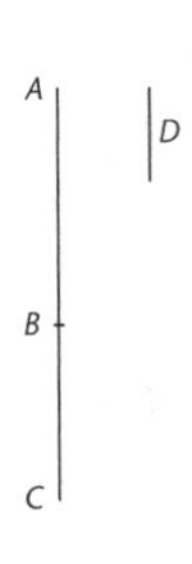

If two incommensurable magnitudes be added together, the whole will also be incommensurable with each of them; and, if the whole be incommensurable with one of them, the original magnitudes will also be incommensurable.

LEMMA. If to any straight line there be applied a parallelogram deficient by a square figure, the applied parallelogram is equal to the rectangle contained by the segments of the straight line resulting from the application.

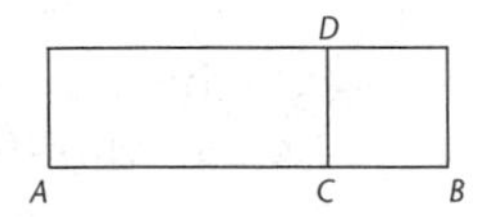

Proposition 17

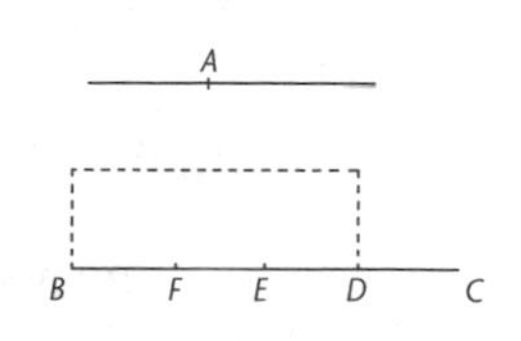

If there be two unequal straight lines, and to the greater there be applied a parallelogram equal to the fourth part of the square on the less and deficient by a square figure, and if it divide it into parts which are commensurable in length, then the square on the greater will be greater than the square on the less by the square on a straight line commensurable with the greater.

And, if the square on the greater be greater than the square on the less by the square on a straight line commensurable with the greater, and if there be applied to the greater a parallelogram equal to the fourth part of the square on the less and deficient by a square figure, it will divide it into parts which are commensurable in length.

Proposition 18

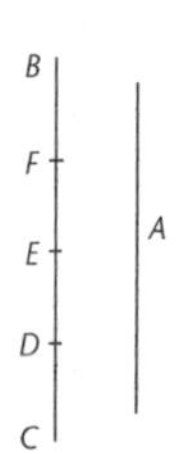

If there be two unequal straight lines, and to the greater there be applied a parallelogram equal to the fourth part of the square on the less and deficient by a square figure, and if it divide it into parts which are incommensurable, the square on the greater will be greater than the square on the less by the square on a straight line incommensurable with the greater.

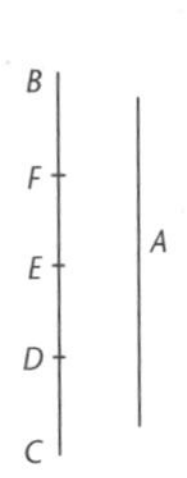

And, if the square on the greater be greater than the square on the less by the square on a straight line incommensurable with the greater, and if there be applied to the greater a parallelogram equal to the fourth part of the square on the less and deficient by a square figure, it divides it into parts which are incommensurable.

⟨**LEMMA.** If any straight line be commensurable in length with a given rational straight line, it is called rational and commensurable with the other not only in length but in square also.

But, if any straight line be commensurable in square with a given rational straight line, then, if it is also commensurable in length with it, it is called in this case also rational and commensurable with it both in length and in square; but, if again any straight line, being commensurable in square with a given rational straight line, be incommensurable in length with it, it is called in this case also rational but commensurable in square only.⟩

Proposition 19

The rectangle contained by rational straight lines commensurable in length is rational.

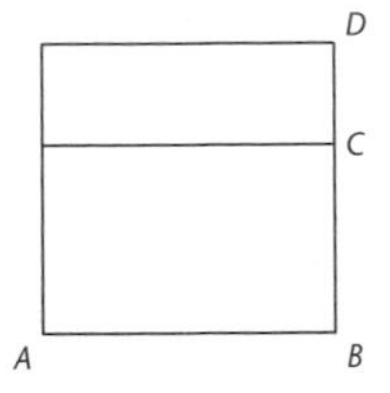

Proposition 20

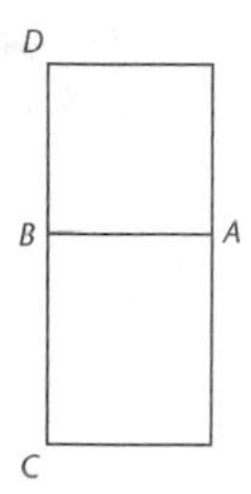

If a rational area be applied to a rational straight line, it produces as breadth a straight line rational and commensurable in length with the straight line to which it is applied.

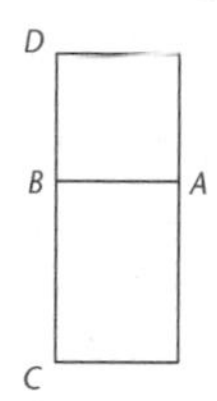

Proposition 21

The rectangle contained by rational straight lines commensurable in square only is irrational, and the side of the square equal to it is irrational. Let the latter be called medial.

LEMMA. If there be two straight lines, then, as the first is to the second, so is the square on the first to the rectangle contained by the two straight lines.

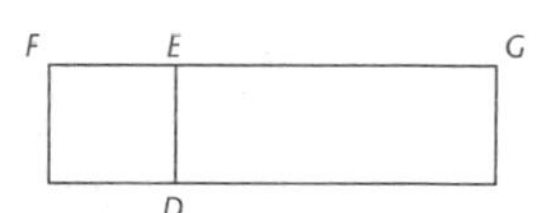

Proposition 22

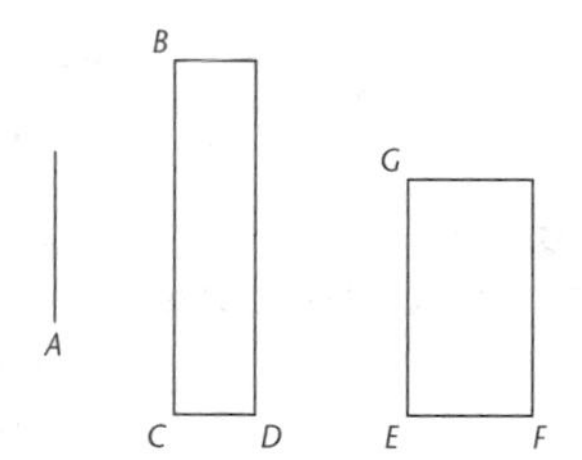

The square on a medial straight line, if applied to a rational straight line, produces as breadth a straight line rational and incommensurable in length with that to which it is applied.

Proposition 23

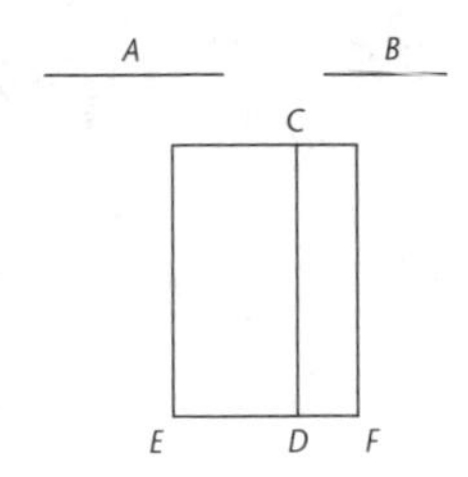

A straight line commensurable with a medial straight line is medial.

PORISM. From this it is manifest that an area commensurable with a medial area is medial.

And in the same way as was explained in the case of rationals [Lemma following X. 18] it follows, as regards medials, that a straight line commensurable in length with a medial straight line is called *medial* and commensurable with it not only in length but in square also, since, in general, straight lines commensurable in length are always commensurable in square also.

But, if any straight line be commensurable in square with a medial straight line, then, if it is also commensurable in length with it, the straight lines are called, in this case too, medial and commensurable in length and in square, but, if in square only, they are called medial straight lines commensurable in square only.

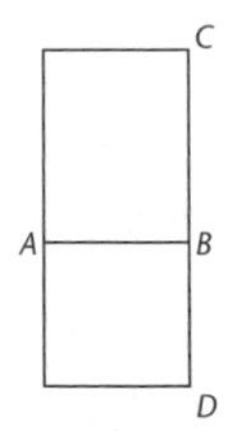

Proposition 24

The rectangle contained by medial straight lines commensurable in length is medial.

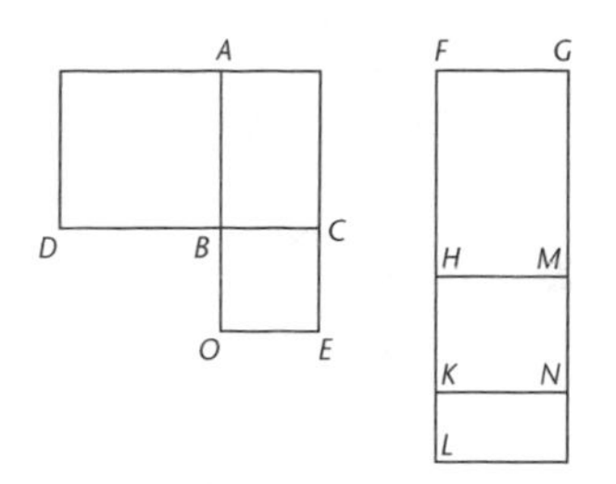

Proposition 25

The rectangle contained by medial straight lines commensurable in square only is either rational or medial.

Proposition 26

A medial area does not exceed a medial area by a rational area.

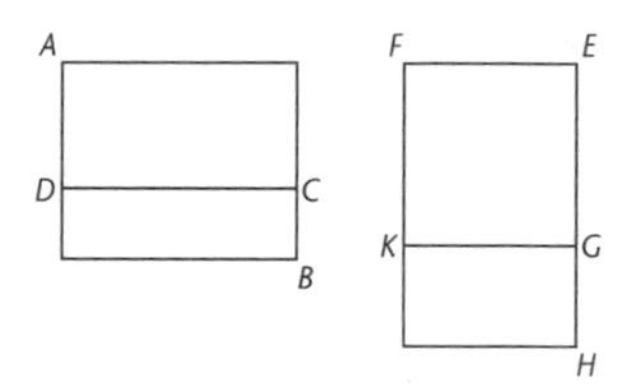

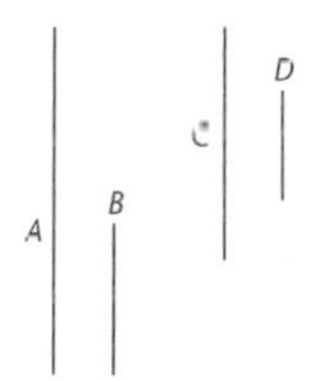

Proposition 27

To find medial straight lines commensurable in square only which contain a rational rectangle.

Proposition 28

To find medial straight lines commensurable in square only, which contain a medial rectangle.

LEMMA 1. To find two square numbers such that their sum is also square.

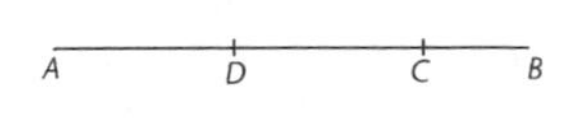

LEMMA 2. To find two square numbers such that their sum is not square.

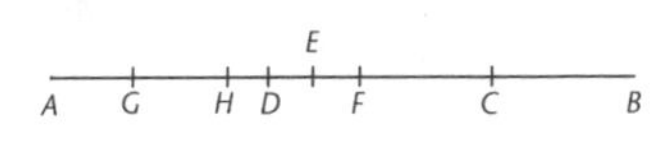

Proposition 29

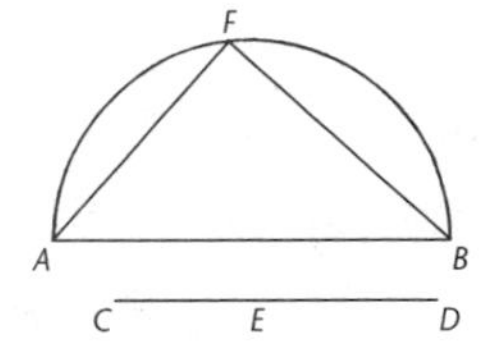

To find two rational straight lines commensurable in square only and such that the square on the greater is greater than the square on the less by the square on a straight line commensurable in length with the greater.

Proposition 30

To find two rational straight lines commensurable in square only and such that the square on the greater is greater than the square on the less by the square on a straight line incommensurable in length with the greater.

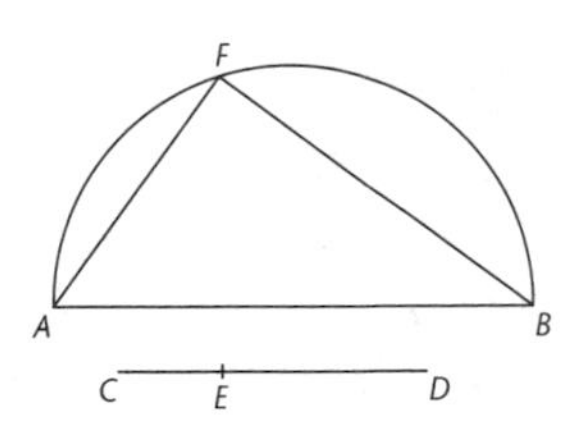

Proposition 31

To find two medial straight lines commensurable in square only, containing a rational rectangle, and such that the square on the greater is greater than the square on the less by the square on a straight line commensurable in length with the greater.

Proposition 32

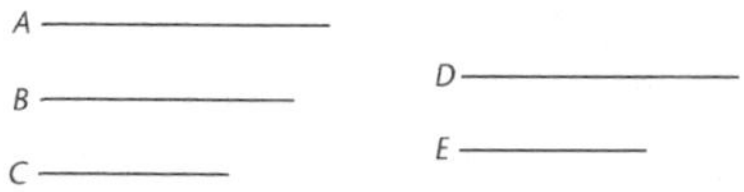

To find two medial straight lines commensurable in square only, containing a medial rectangle, and such that the square on the greater is greater than the square on the less by the square on a straight line commensurable with the greater.

LEMMA. The rectangle *BC, AD* is equal to the rectangle *BA, AC.*

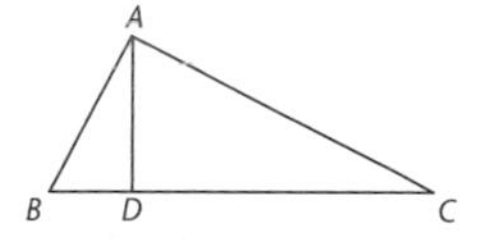

Proposition 33

To find two straight lines incommensurable in square which make the sum of the squares on them rational but the rectangle contained by them medial.

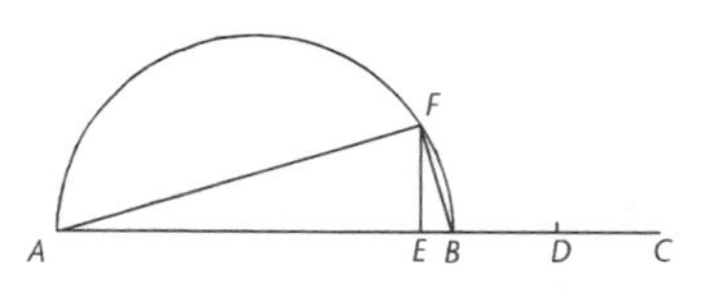

Proposition 34

To find two straight lines incommensurable in square which make the sum of the squares on them medial but the rectangle contained by them rational.

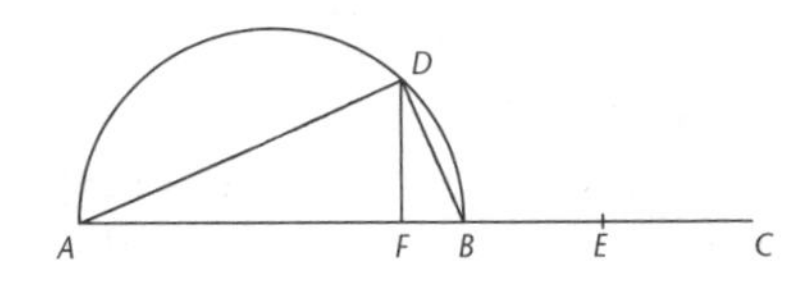

Proposition 35

To find two straight lines incommensurable in square which make the sum of the squares on them medial and the rectangle contained by them medial and moreover incommensurable with the sum of the squares on them.

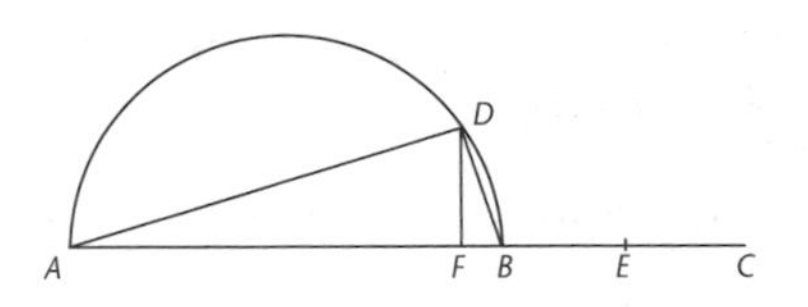

Proposition 36

If two rational straight lines commensurable in square only be added together, the whole is irrational; and let it be called binomial.

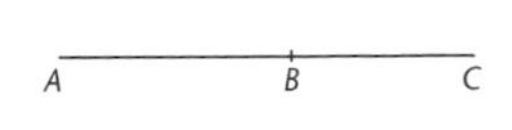

Proposition 37

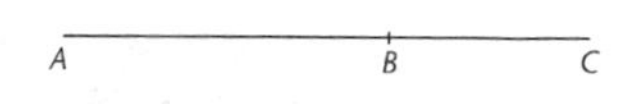

If two medial straight lines commensurable in square only and containing a rational rectangle be added together, the whole is irrational; and let it be called a first bimedial straight line.

Proposition 38

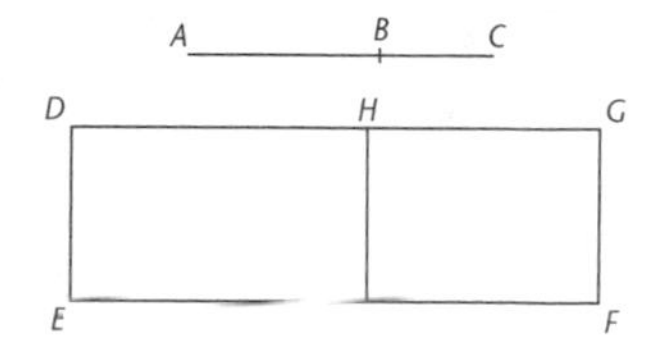

If two medial straight lines commensurable in square only and containing a medial rectangle be added together, the whole is irrational; and let it be called a second bimedial straight line.

Proposition 39

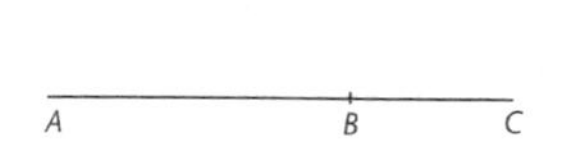

If two straight lines incommensurable in square which make the sum of the squares on them rational, but the rectangle contained by them medial, be added together, the whole straight line is irrational: and let it be called major.

Proposition 40

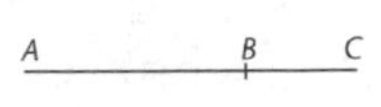

If two straight lines incommensurable in square which make the sum of the squares on them medial, but the rectangle contained by them rational, be added together, the whole straight line is irrational; and let it be called the side of a rational plus a medial area.

Proposition 41

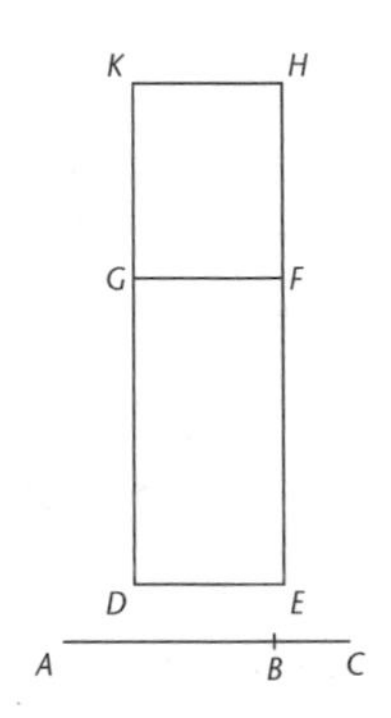

If two straight lines incommensurable in square which make the sum of the squares on them medial, and the rectangle contained by them medial and also incommensurable with the sum of the squares on them, be added together, the whole straight line is irrational; and let it be called the side of the sum of two medial areas.

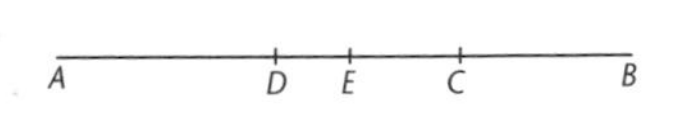

LEMMA. The aforesaid irrational straight lines are divided only in one way into the straight lines of which they are the sum and which produce the types in question.

Proposition 42

A binomial straight line is divided into its terms at one point only.

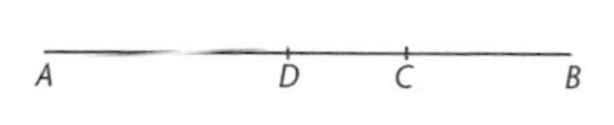

Proposition 43

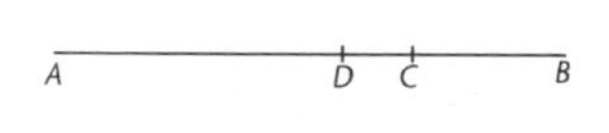

A first bimedial straight line is divided at one point only.

Proposition 44

A second bimedial straight line is divided at one point only.

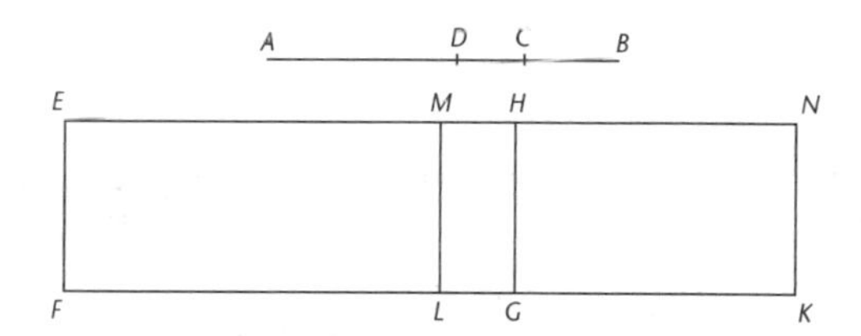

Proposition 45

A major straight line is divided at one and the same point only.

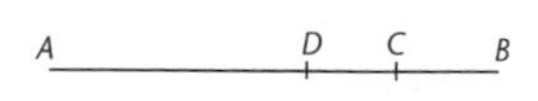

Proposition 46

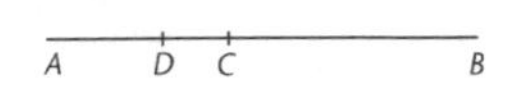

The side of a rational plus a medial area is divided at one point only.

Proposition 47

The side of the sum of two medial areas is divided at one point only.

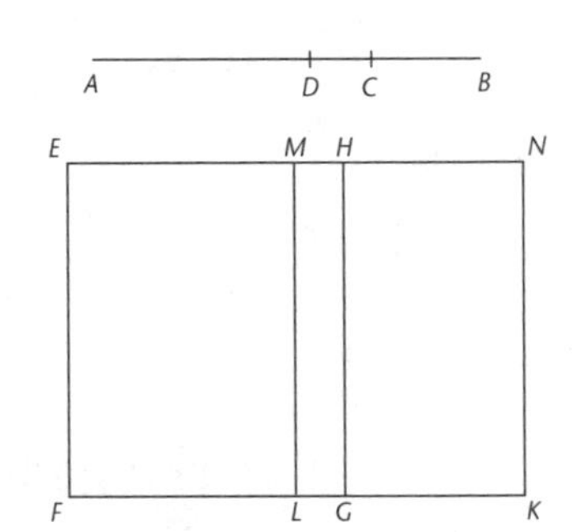

Definitions II

1. Given a rational straight line and a binomial, divided into its terms, such that the square on the greater term is greater than the square on the lesser by the square on a straight line commensurable in length with the greater, then, if the greater term be commensurable in length with the rational straight line set out, let the whole be called a *first binomial* straight line;

2. but if the lesser term be commensurable in length with the rational straight line set out, let the whole be called a *second binomial*;

3. and if neither of the terms be commensurable in length with the rational straight line set out, let the whole be called a *third binomial.*
4. Again, if the square on the greater term be greater than the square on the lesser by the square on a straight line incommensurable in length with the greater, then, if the greater term be commensurable in length with the rational straight line set out, let the whole be called a *fourth binomial*;
5. if the lesser, a *fifth binomial*;
6. and if neither, a *sixth binomial.*

Proposition 48

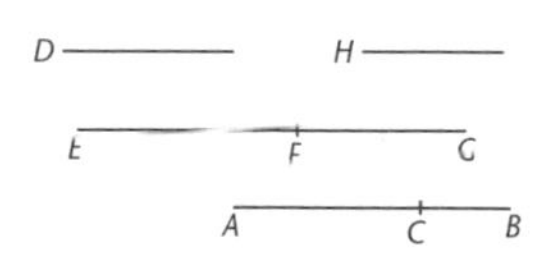

To find the first binomial straight line.

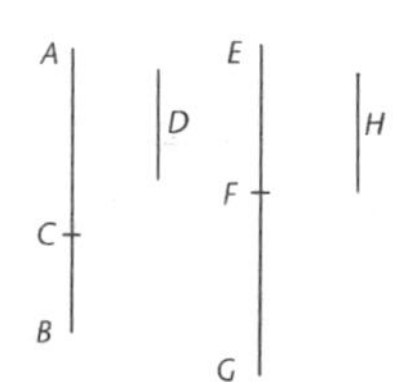

Proposition 49

To find the second binomial straight line.

Proposition 50

To find the third binomial straight line.

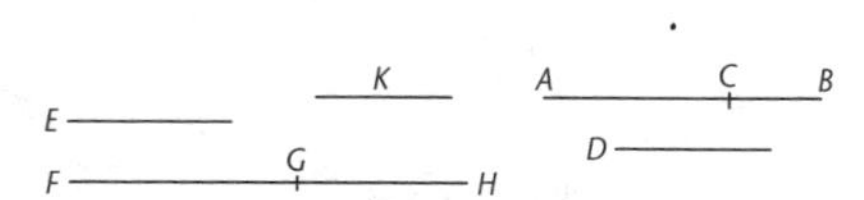

Proposition 51

To find the fourth binomial straight line.

Proposition 52

To find the fifth binomial straight line.

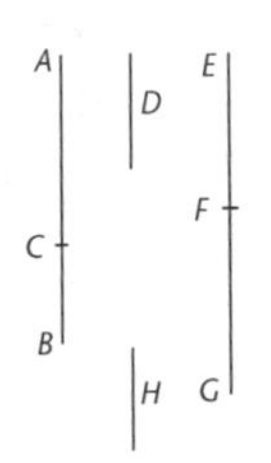

Proposition 53

To find the sixth binomial straight line.

LEMMA. Let there be two squares *AB, BC,* and let them be placed so that *DB* is in a straight line with *BE*; …

Let the parallelogram *AC* be completed;
I say that *AC* is a square, that *DG* is a mean proportional between *AB, BC,* and further that *DC* is a mean proportional between *AC, CB.*

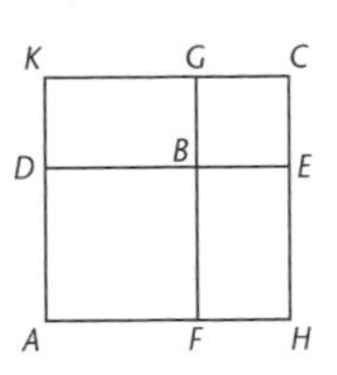

Proposition 54

If an area be contained by a rational straight line and the first binomial, the "side" of the area is the irrational straight line which is called binomial.

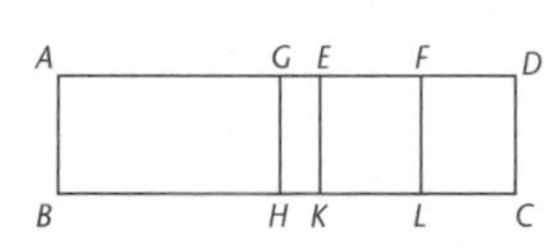

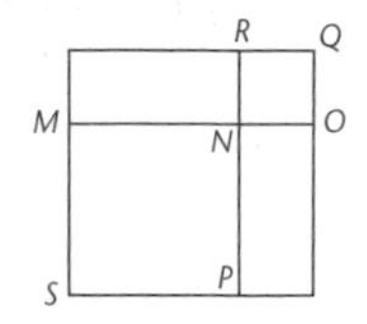

Proposition 55

If an area be contained by a rational straight line and the second binomial, the "side" of the area is the irrational straight line which is called a first bimedial.

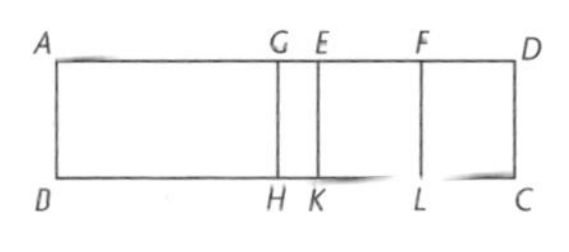

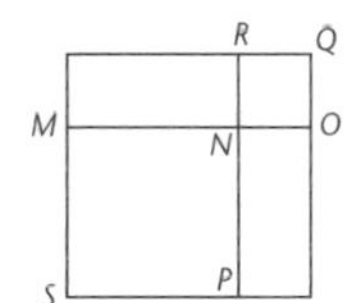

Proposition 56

If an area be contained by a rational straight line and the third binomial, the "side" of the area is the irrational straight line called a second bimedial.

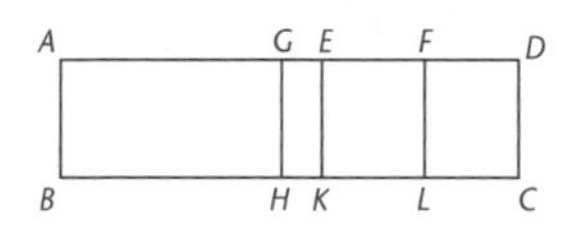

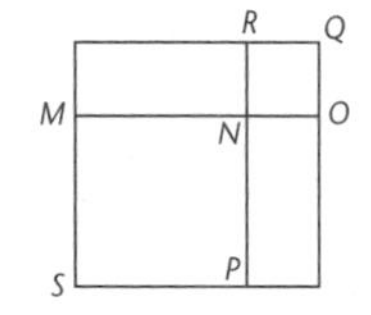

Proposition 57

If an area be contained by a rational straight line and the fourth binomial, the "side" of the area is the irrational straight line called major.

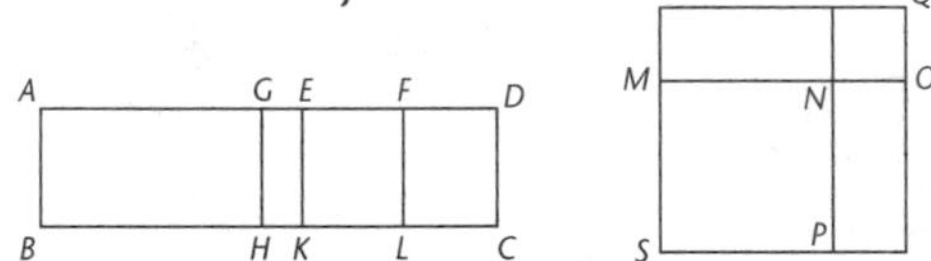

Proposition 58

If an area be contained by a rational straight line and the fifth binomial, the "side" of the area is the irrational straight line called the side of a rational plus a medial area.

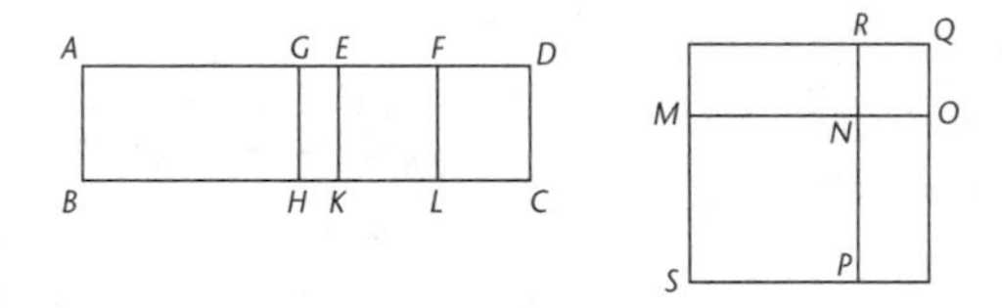

Proposition 59

If an area be contained by a rational straight line and the sixth binomial, the "side" of the area is the irrational straight line called the side of the sum of two medial areas.

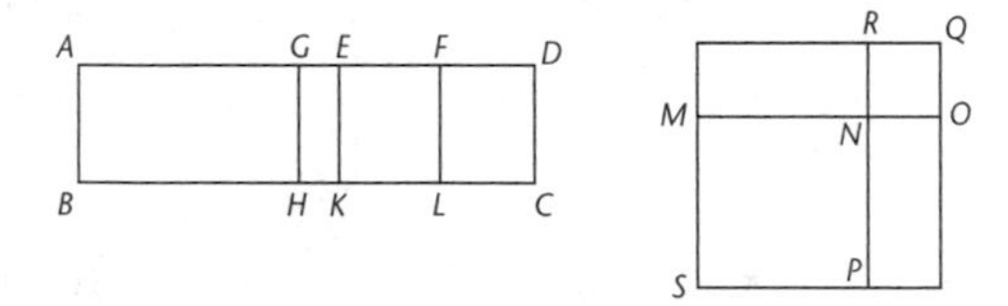

⟨**LEMMA.** If a straight line be cut into unequal parts, the squares on the unequal parts are greater than twice the rectangle contained by the unequal parts.⟩

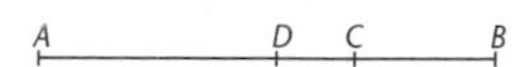

Proposition 60

The square on the binomial straight line applied to a rational straight line produces as breadth the first binomial.

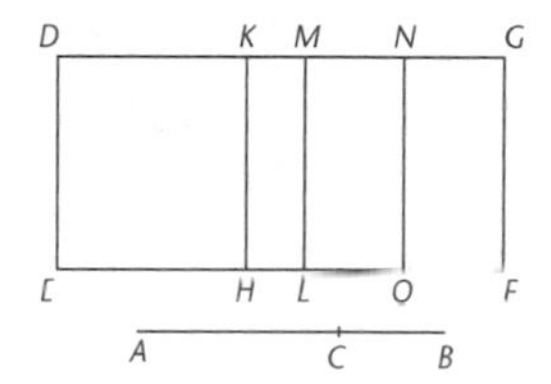

Proposition 61

The square on the first bimedial straight line applied to a rational straight line produces as breadth the second binomial.

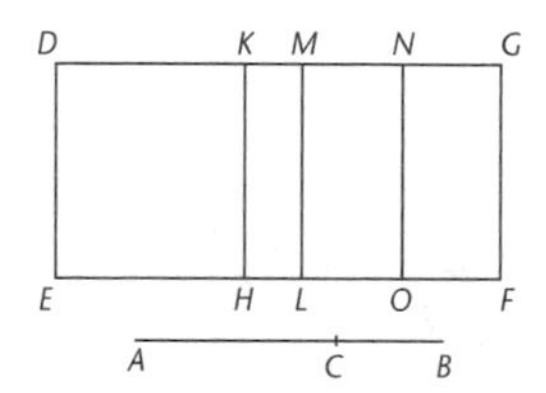

Proposition 62

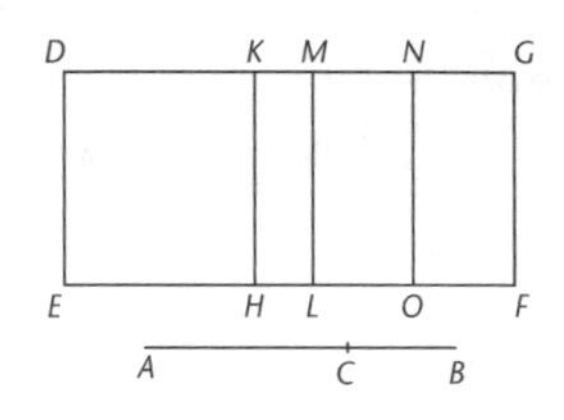

The square on the second bimedial straight line applied to a rational straight line produces as breadth the third binomial.

Proposition 63

The square on the major straight line applied to a rational straight line produces as breadth the fourth binomial.

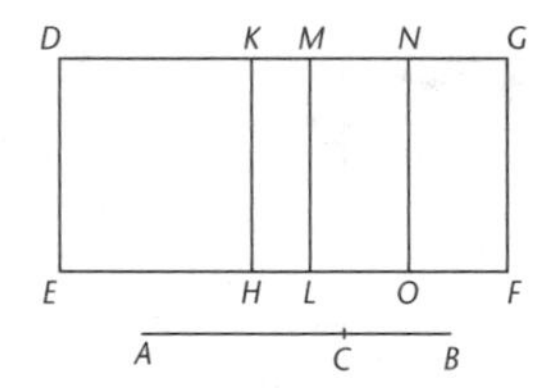

Proposition 64

The square on the side of a rational plus a medial area applied to a rational straight line produces as breadth the fifth binomial.

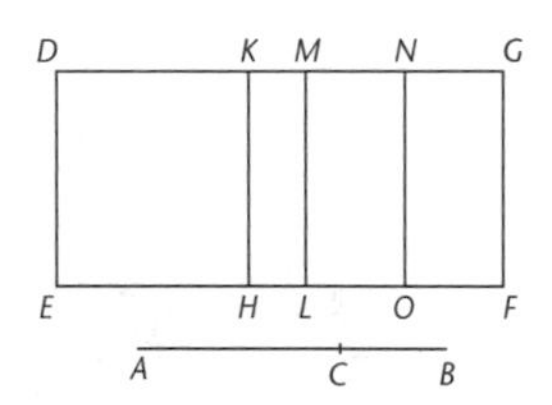

Proposition 65

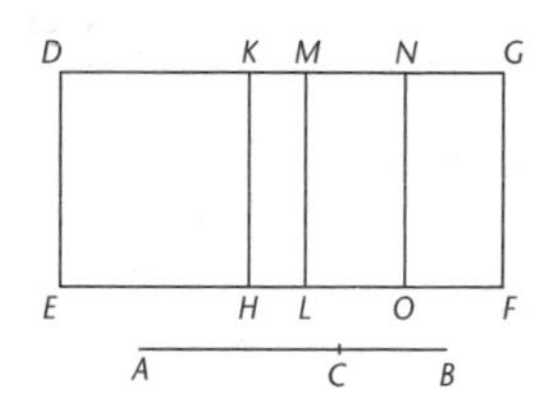

The square on the side of the sum of two medial areas applied to a rational straight line produces as breadth the sixth binomial.

Proposition 66

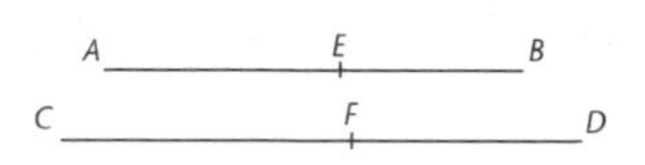

A straight line commensurable in length with a binomial straight line is itself also binomial and the same in order.

Proposition 67

A straight line commensurable in length with a bimedial straight line is itself also bimedial and the same in order.

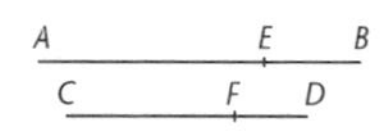

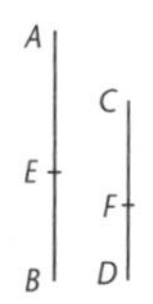

Proposition 68

A straight line commensurable with a major straight line is itself also major.

Proposition 69

A straight line commensurable with the side of a rational plus a medial area is itself also the side of a rational plus a medial area.

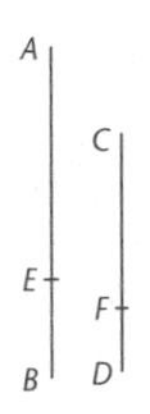

Proposition 70

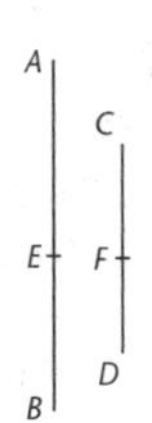

A straight line commensurable with the side of the sum of two medial areas is the side of the sum of two medial areas.

Proposition 71

If a rational and a medial area be added together, four irrational straight lines arise, namely a binomial or a first bimedial or a major or a side of a rational plus a medial area.

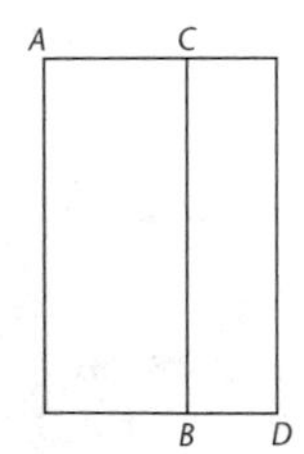

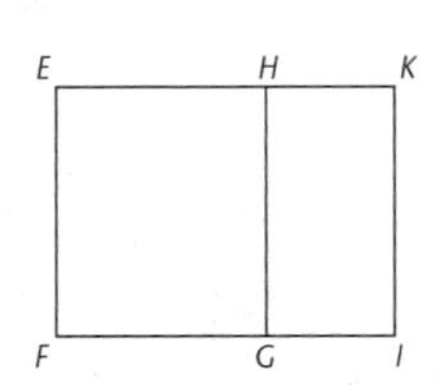

Proposition 72

If two medial areas incommensurable with one another be added together, the remaining two irrational straight lines arise, namely either a second bimedial or a side of the sum of two medial areas.

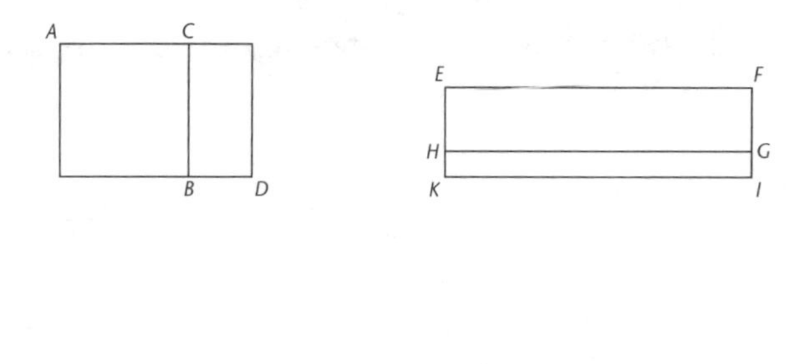

The binomial straight line and the irrational straight lines after it are neither the same with the medial nor with one another.

Proposition 73

If from a rational straight line there be subtracted a rational straight line commensurable with the whole in square only, the remainder is irrational; and let it be called an apotome.

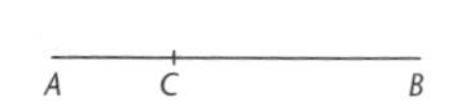

Proposition 74

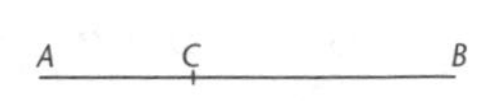

If from a medial straight line there be subtracted a medial straight line which is commensurable with the whole in square only, and which contains with the whole a rational rectangle, the remainder is irrational. And let it be called a first apotome of a medial straight line.

Proposition 75

If from a medial straight line there be subtracted a medial straight line which is commensurable with the whole in square only, and which contains with the whole a medial rectangle, the remainder is irrational; and let it be called a second apotome of a medial straight line.

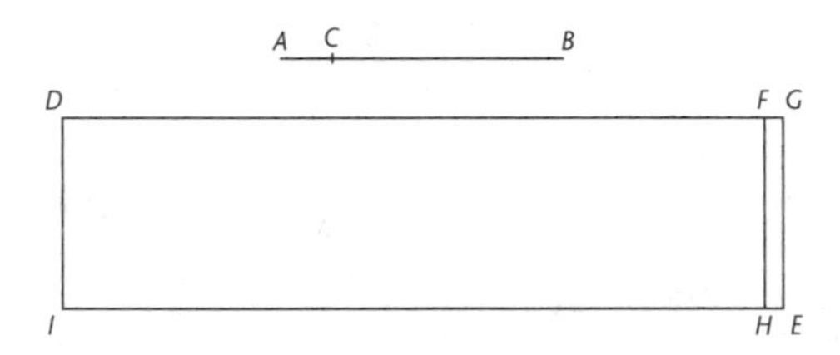

Proposition 76

If from a straight line there be subtracted a straight line which is incommensurable in square with the whole and which with the whole makes the squares on them added together rational, but the rectangle contained by them medial, the remainder is irrational; and let it be called minor.

Proposition 77

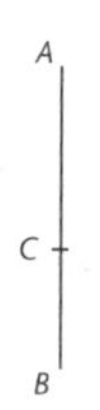

If from a straight line there be subtracted a straight line which is incommensurable in square with the whole, and which with the whole makes the sum of the squares on them medial, but twice the rectangle contained by them rational, the remainder is irrational; and let it be called that which produces with a rational area a medial whole.

Proposition 78

If from a straight line there be subtracted a straight line which is incommensurable in square with the whole and which with the whole makes the sum of the squares on them medial, twice the rectangle contained by them medial, and further, the squares on them incommensurable with twice the rectangle contained by them, the remainder is irrational; and let it be called that which produces with a medial area a medial whole.

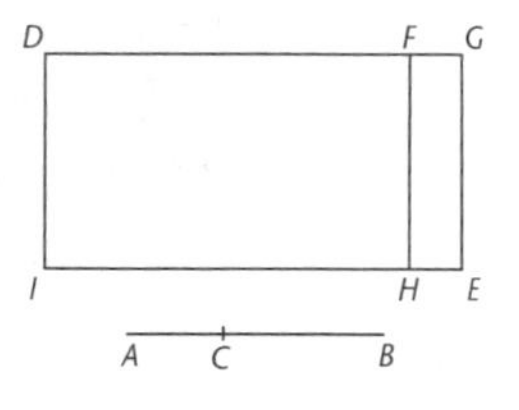

Proposition 79

To an apotome only one rational straight line can be annexed which is commensurable with the whole in square only.

Proposition 80

To a first apotome of a medial straight line only one medial straight line can be annexed which is commensurable with the whole in square only and which contains with the whole a rational rectangle.

Proposition 81

To a second apotome of a medial straight line only one medial straight line can be annexed which is commensurable with the whole in square only and which contains with the whole a medial rectangle.

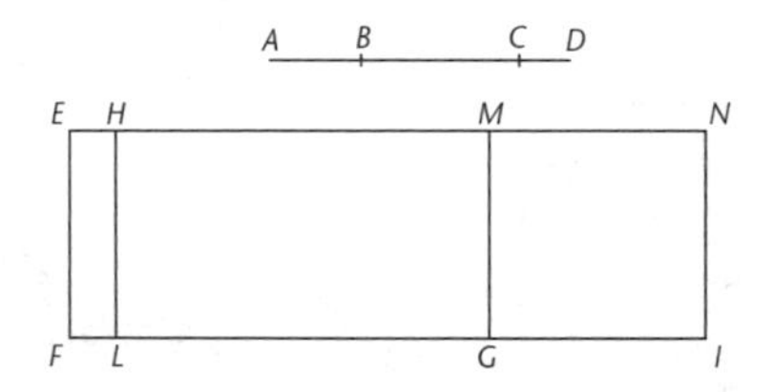

Proposition 82

To a minor straight line only one straight line can be annexed which is incommensurable in square with the whole and which makes, with the whole, the sum of the squares on them rational but twice the rectangle contained by them medial.

Proposition 83

To a straight line which produces with a rational area a medial whole only one straight line can be annexed which is incommensurable in square with the whole straight line and which with the whole straight line makes the sum of the squares on them medial, but twice the rectangle contained by them rational.

Proposition 84

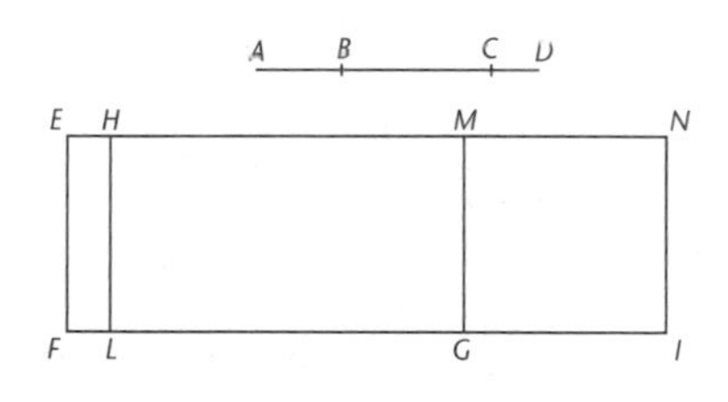

To a straight line which produces with a medial area a medial whole only one straight line can be annexed which is incommensurable in square with the whole straight line and which with the whole straight line makes the sum of the squares on them medial and twice the rectangle contained by them both medial and also incommensurable with the sum of the squares on them.

Definitions III

1. Given a rational straight line and an apotome, if the square on the whole be greater than the square on the annex by the square on a straight line commensurable in length with the whole, and the whole be commensurable in length with the rational straight line set out, let the apotome be called a *first apotome.*

2. But if the annex be commensurable in length with the rational straight line set out, and the square on the whole be greater than that on the annex by the square on a straight line commensurable with the whole, let the apotome be called a *second apotome.*

3. But if neither be commensurable in length with the rational straight line set out, and the square on the whole be greater than the square on the annex by the square on a straight line commensurable with the whole, let the apotome be called a *third apotome.*

4. Again, if the square on the whole be greater than the square on the annex by the square on a straight line incommensurable with the whole, then, if the whole be commensurable in length with the rational straight line set out, let the apotome be called a *fourth apotome*;

5. if the annex be so commensurable, a *fifth*;

6. and, if neither, a *sixth.*

Proposition 85

To find the first apotome.

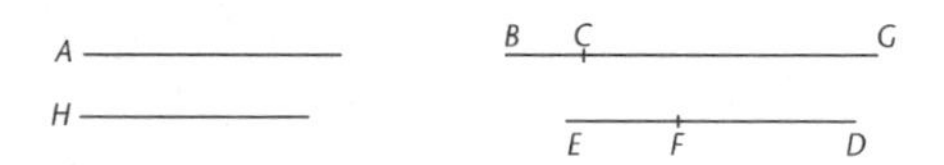

Proposition 86

To find the second apotome.

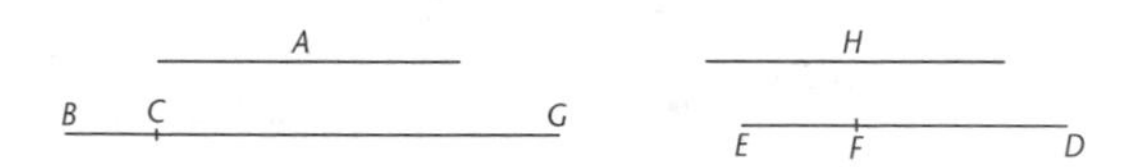

Proposition 87

To find the third apotome.

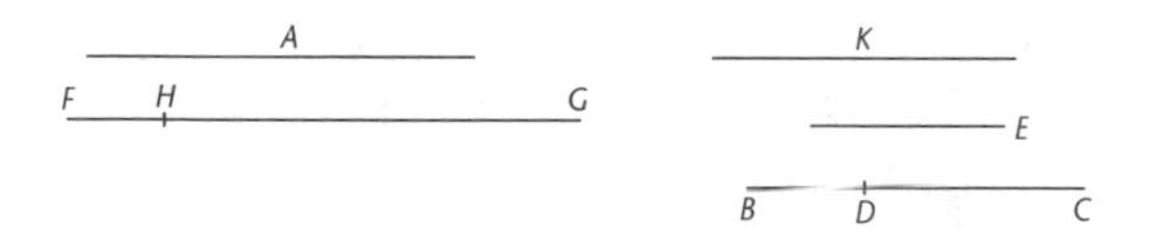

Proposition 88

To find the fourth apotome.

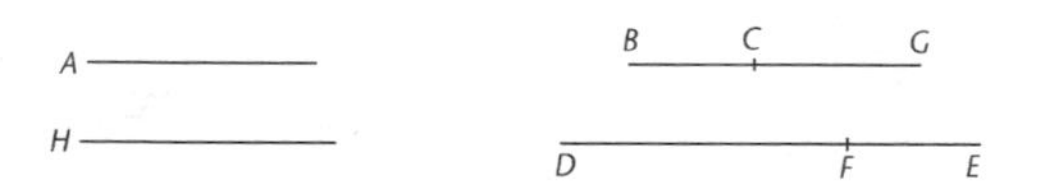

Proposition 89

To find the fifth apotome.

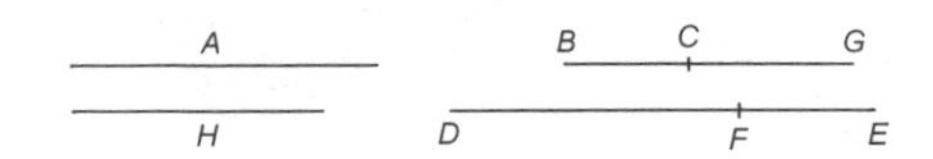

Proposition 90

To find the sixth apotome.

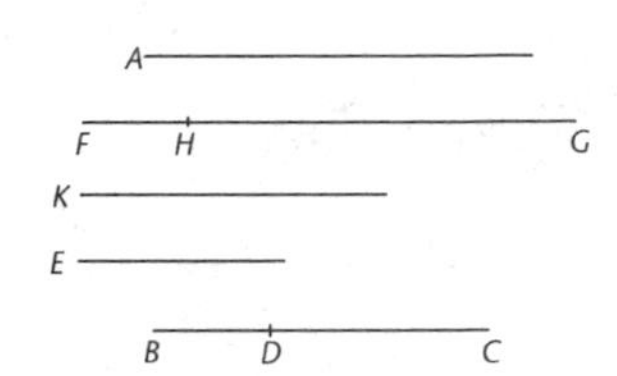

Proposition 91

If an area be contained by a rational straight line and a first apotome, the "side" of the area is an apotome.

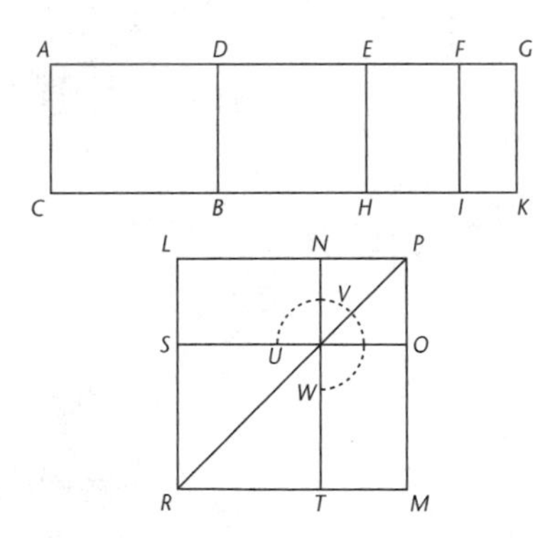

Proposition 92

If an area be contained by a rational straight line and a second apotome, the "side" of the area is a first apotome of a medial straight line.

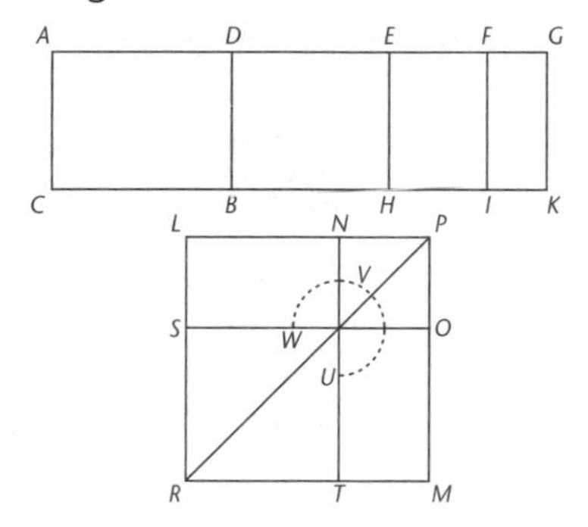

Proposition 93

If an area be contained by a rational straight line and a third apotome, the "side" of the area is a second apotome of a medial straight line.

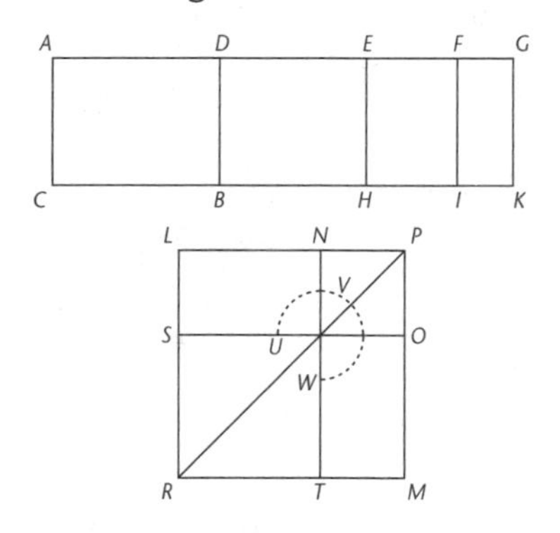

Proposition 94

If an area be contained by a rational straight line and a fourth apotome, the "side" of the area is minor.

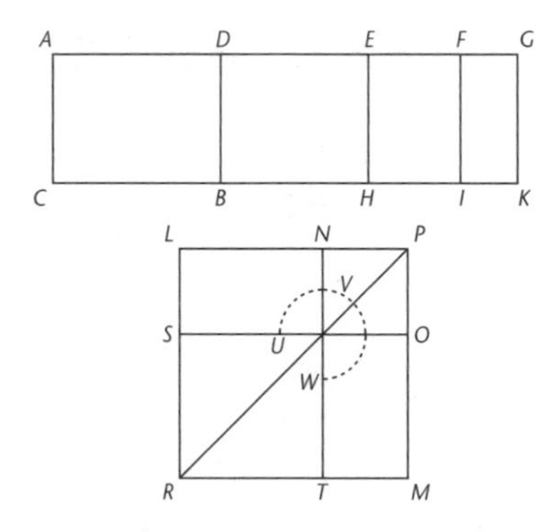

Proposition 95

If an area be contained by a rational straight line and a fifth apotome, the "side" of the area is a straight line which produces with a rational area a medial whole.

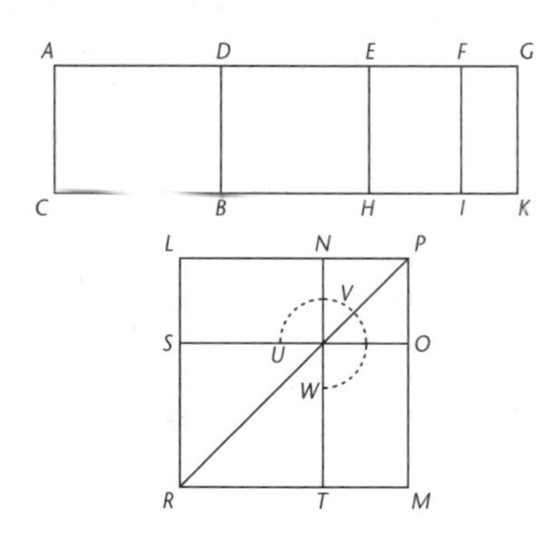

Proposition 96

If an area be contained by a rational straight line and a sixth apotome, the "side" of the area is a straight line which produces with a medial area a medial whole.

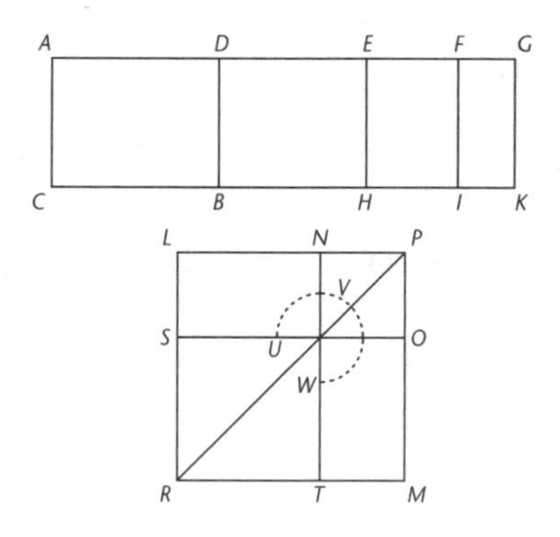

Proposition 97

The square on an apotome applied to a rational straight line produces as breadth a first apotome.

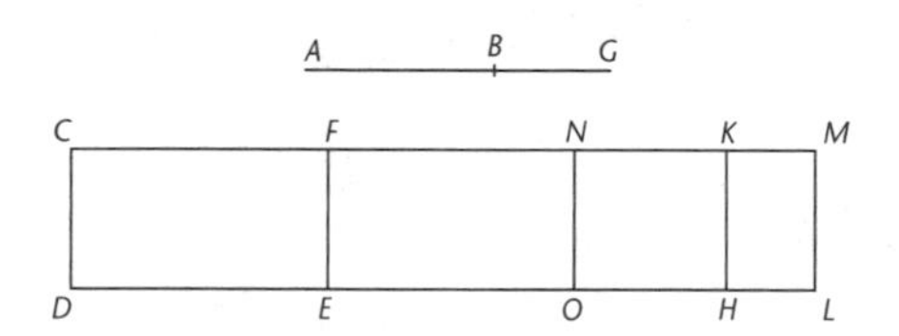

Proposition 98

The square on a first apotome of a medial straight line applied to a rational straight line produces as breadth a second apotome.

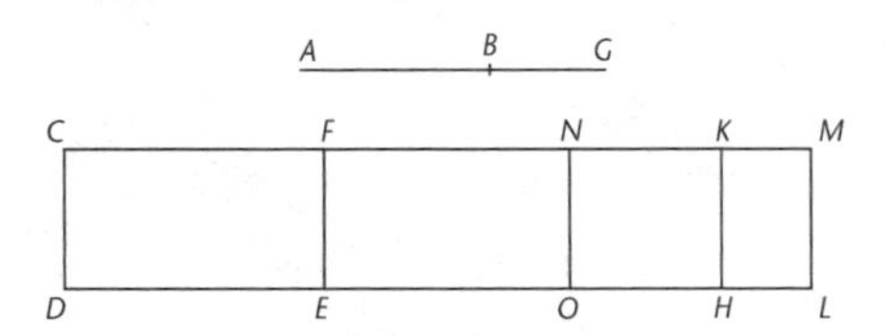

Proposition 99

The square on a second apotome of a medial straight line applied to a rational straight line produces as breadth a third apotome.

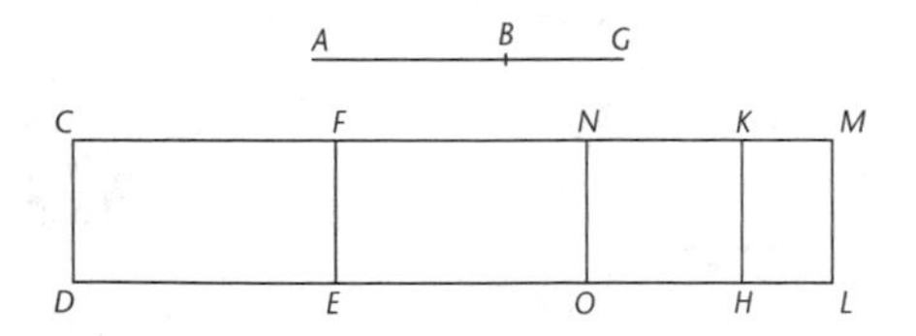

Proposition 100

The square on a minor straight line applied to a rational straight line produces as breadth a fourth apotome.

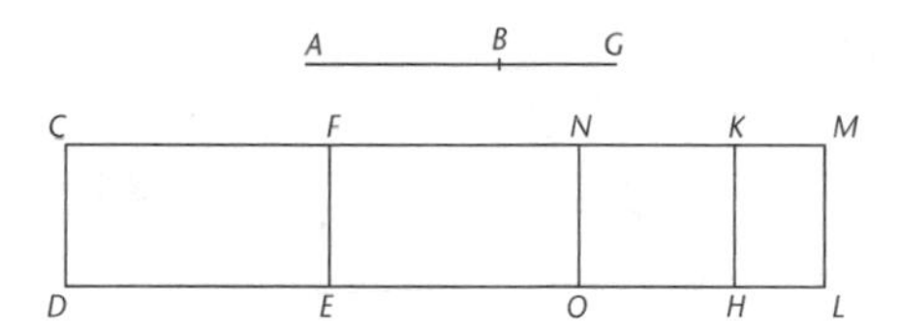

Proposition 101

The square on the straight line which produces with a rational area a medial whole, if applied to a rational straight line, produces as breadth a fifth apotome.

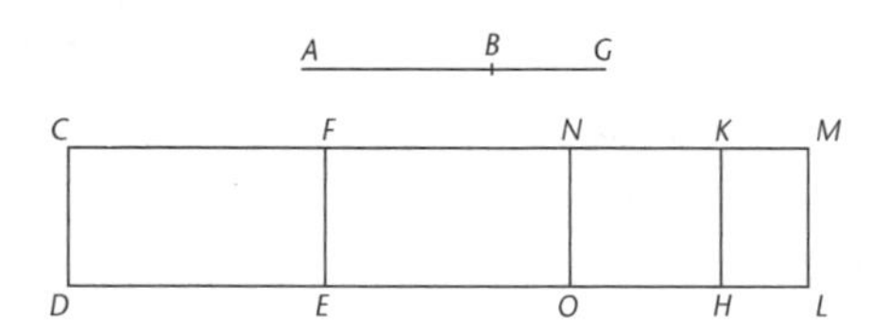

Proposition 102

The square on the straight line which produces with a medial area a medial whole, if applied to a rational straight line, produces as breadth a sixth apotome.

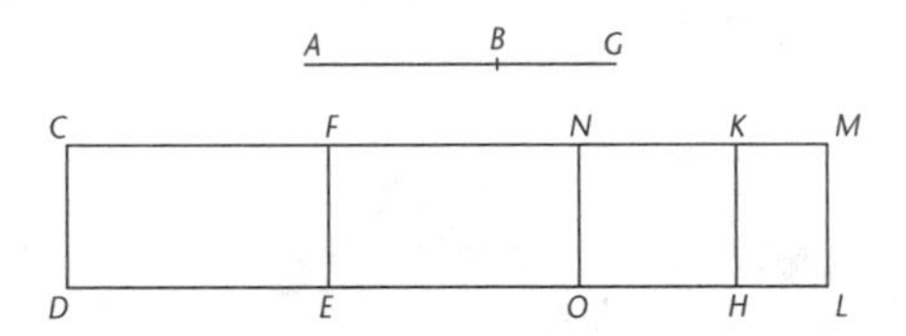

Proposition 103

A straight line commensurable in length with an apotome is an apotome and the same in order.

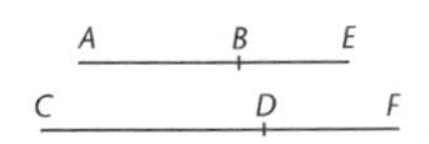

Proposition 104

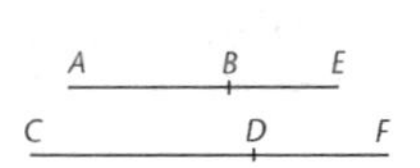

A straight line commensurable with an apotome of a medial straight line is an apotome of a medial straight line and the same in order.

Proposition 105

A straight line commensurable with a minor straight line is minor.

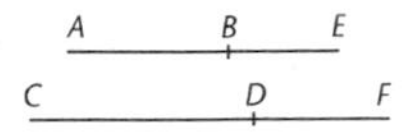

Proposition 106

A straight line commensurable with that which produces with a rational area a medial whole is a straight line which produces with a rational area a medial whole.

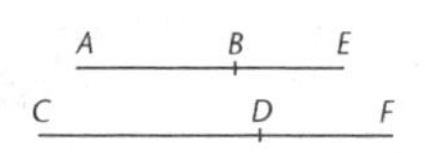

Proposition 107

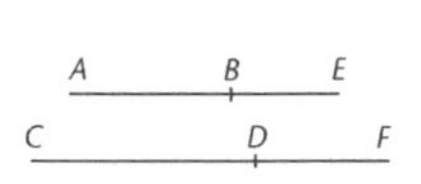

A straight line commensurable with that which produces with a medial area a medial whole is itself also a straight line which produces with a medial area a medial whole.

Proposition 108

If from a rational area a medial area be subtracted, the "side" of the remaining area becomes one of two irrational straight lines, either an apotome or a minor straight line.

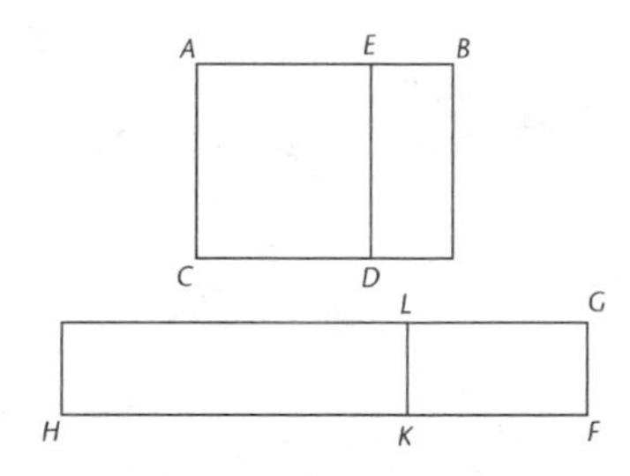

Proposition 109

If from a medial area a rational area be subtracted, there arise two other irrational straight lines, either a first apotome of a medial straight line or a straight line which produces with a rational area a medial whole.

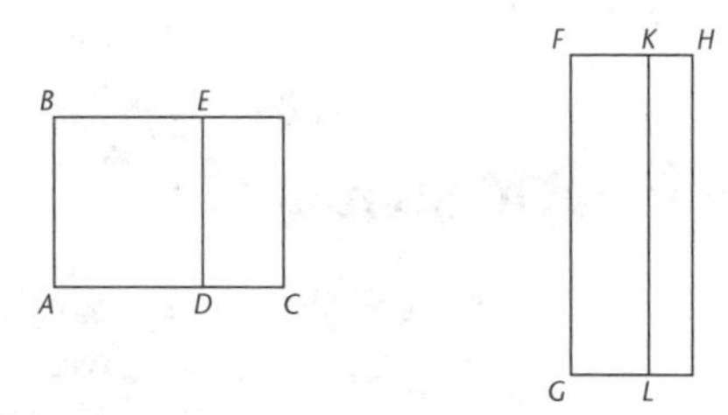

Proposition 110

If from a medial area there be subtracted a medial area incommensurable with the whole, the two remaining irrational straight lines arise, either a second apotome of a medial straight line or a straight line which produces with a medial area a medial whole.

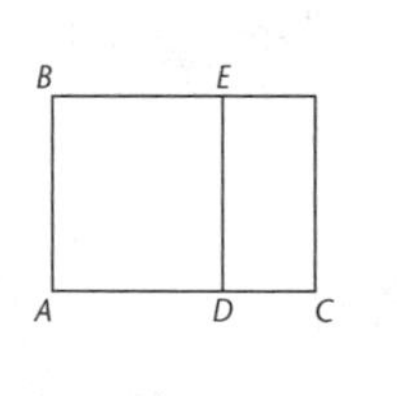

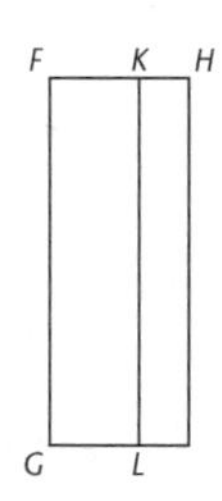

Proposition 111

The apotome is not the same with the binomial straight line.

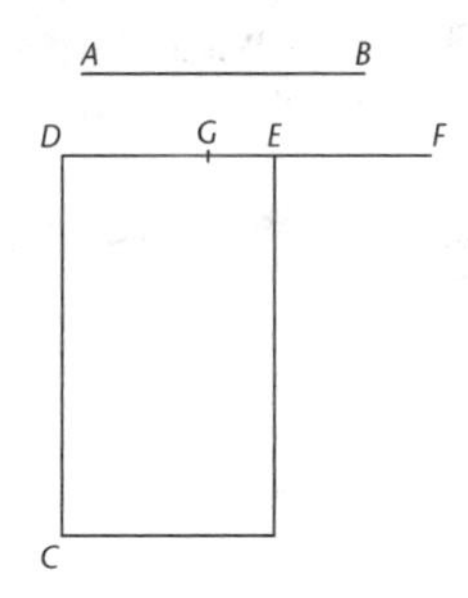

The apotome and the irrational straight lines following it are neither the same with the medial straight line nor with one another.

...[T]here are, in order, thirteen irrational straight lines in all,

Medial,
Binomial,
First bimedial,
Second bimedial,
Major,
"Side" of a rational plus a medial area,
"Side" of the sum of two medial areas,
Apotome,
First apotome of a medial straight line,
Second apotome of a medial straight line,
Minor,
Producing with a rational area a medial whole,
Producing with a medial area a medial whole.

Proposition 112

The square on a rational straight line applied to the binomial straight line produces as breadth an apotome the terms of which are commensurable with the terms of the binomial and moreover in the same ratio; and further, the apotome so arising will have the same order as the binomial straight line.

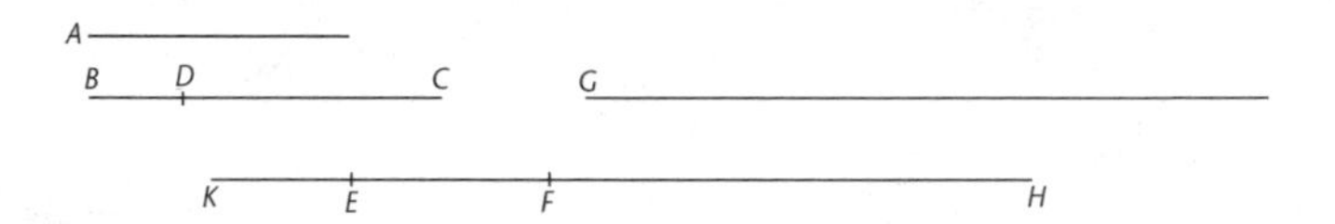

Proposition 113

The square on a rational straight line, if applied to an apotome, produces as breadth the binomial straight line the terms of which are commensurable with the terms of the apotome and in the same ratio; and further, the binomial so arising has the same order as the apotome.

Proposition 114

If an area be contained by an apotome and the binomial straight line the terms of which are commensurable with the terms of the apotome and in the same ratio, the "side" of the area is rational.

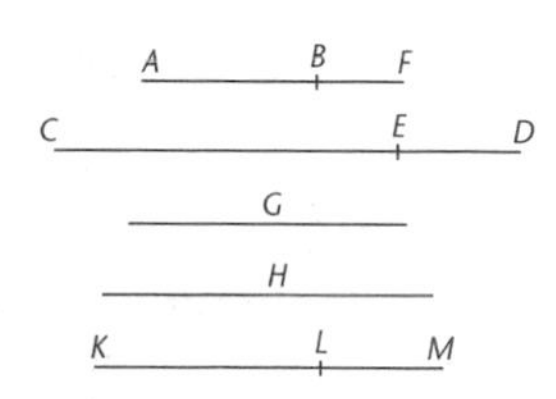

PORISM. And it is made manifest to us by this also that it is possible for a rational area to be contained by irrational straight lines.

Proposition 115

From a medial straight line there arise irrational straight lines infinite in number, and none of them is the same as any of the preceding.

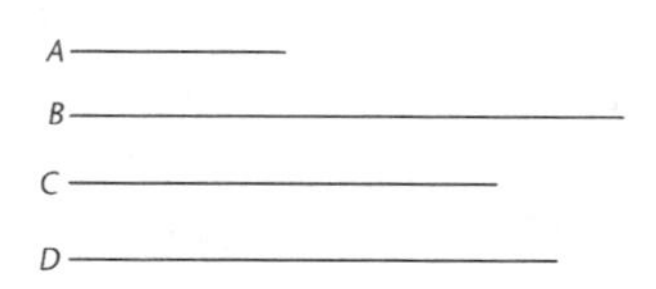

Book XI

Definitions

1. A *solid* is that which has length, breadth, and depth.
2. An extremity of a solid is a surface.
3. A *straight line* is at *right angles to a plane,* when it makes right angles with all the straight lines which meet it and are in the plane.
4. A *plane* is *at right angles to a plane* when the straight lines drawn, in one of the planes, at right angles to the common section of the planes, are at right angles to the remaining plane.
5. The *inclination of a straight line to a plane* is, assuming a perpendicular drawn from the extremity of the straight line which is elevated above the plane to the plane, and a straight line joined from the point thus arising to the extremity of the straight line which is in the plane, the angle contained by the straight line so drawn and the straight line standing up.
6. The *inclination of a plane to a plane* is the acute angle contained by the straight lines drawn at right angles to the common section at the same point, one in each of the planes.
7. A plane is said to be *similarly inclined* to a plane as another is to another when the said angles of the inclinations are equal to one another.

8. *Parallel planes* are those which do not meet.

9. *Similar solid figures* are those contained by similar planes equal in multitude.

10. *Equal and similar solid figures* are those contained by similar planes equal in multitude and in magnitude.

11. A *solid angle* is the inclination constituted by more than two lines which meet one another and are not in the same surface, towards all the lines.

    Otherwise: A *solid angle* is that which is contained by more than two plane angles which are not in the same plane and are constructed to one point.

12. A *pyramid* is a solid figure, contained by planes, which is constructed from one plane to one point.

13. A *prism* is a solid figure contained by planes two of which, namely those which are opposite, are equal, similar and parallel, while the rest are parallelograms.

14. When, the diameter of a semicircle remaining fixed, the semicircle is carried round and restored again to the same position from which it began to be moved, the figure so comprehended is a *sphere.*

15. The *axis of the sphere* is the straight line which remains fixed and about which the semicircle is turned.

16. The *centre of the sphere* is the same as that of the semicircle.

17. A *diameter of the sphere* is any straight line drawn through the centre and terminated in both directions by the surface of the sphere.

18. When, one side of those about the right angle in a right-angled triangle remaining fixed, the triangle is carried round and restored again to the same position from which it began to be moved, the figure so comprehended is a *cone.*

    And, if the straight line which remains fixed be equal to the remaining side about the right angle which is carried round, the cone will be *right-angled;* if less, *obtuse-angled;* and if greater, *acute-angled.*

19. The *axis of the cone* is the straight line which remains fixed and about which the triangle is turned.

20. And the *base* is the circle described by the straight line which is carried round.

21. When, one side of those about the right angle in a rectangular parallelogram remaining fixed, the parallelogram is carried round and restored again to the same position from which it began to be moved, the figure so comprehended is a *cylinder.*

22. The *axis of the cylinder* is the straight line which remains fixed and about which the parallelogram is turned.

23. And the *bases* are the circles described by the two sides opposite to one another which are carried round.

24. *Similar cones and cylinders* are those in which the axes and the diameters of the bases are proportional.

25. A *cube* is a solid figure contained by six equal squares.

26. An *octahedron* is a solid figure contained by eight equal and equilateral triangles.
27. An *icosahedron* is a solid figure contained by twenty equal and equilateral triangles.
28. A *dodecahedron* is a solid figure contained by twelve equal, equilateral, and equiangular pentagons.

Proposition 1

A part of a straight line cannot be in the plane of reference and a part in a plane more elevated.

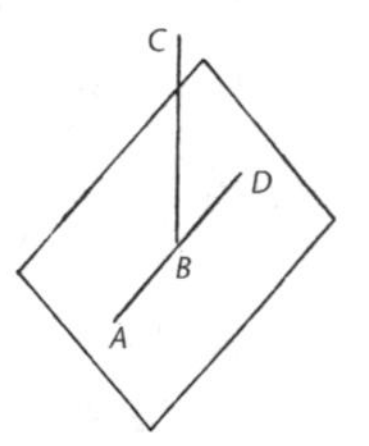

Proposition 2

If two straight lines cut one another, they are in one plane, and every triangle is in one plane.

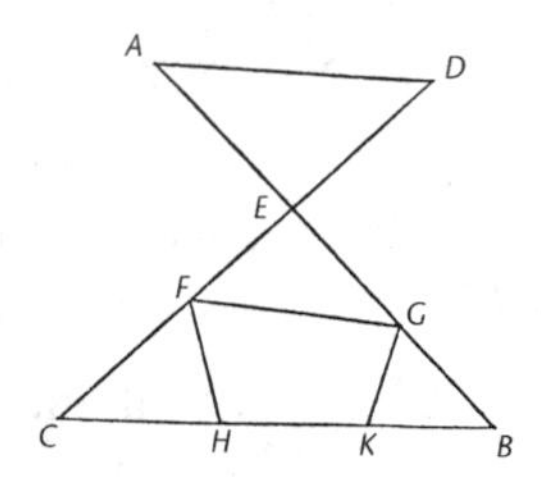

Proposition 3

If two planes cut one another, their common section is a straight line.

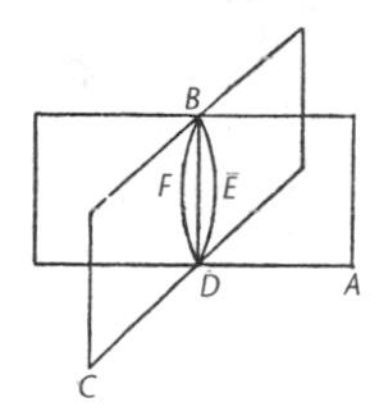

Proposition 4

If a straight line be set up at right angles to two straight lines which cut one another, at their common point of section, it will also be at right angles to the plane through them.

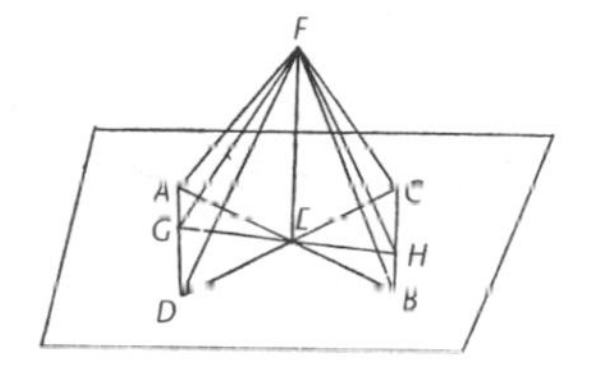

Proposition 5

If a straight line be set up at right angles to three straight lines which meet one another, at their common point of section, the three straight lines are in one plane.

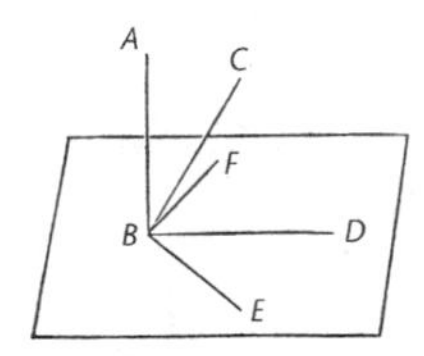

Proposition 6

If two straight lines be at right angles to the same plane, the straight lines will be parallel.

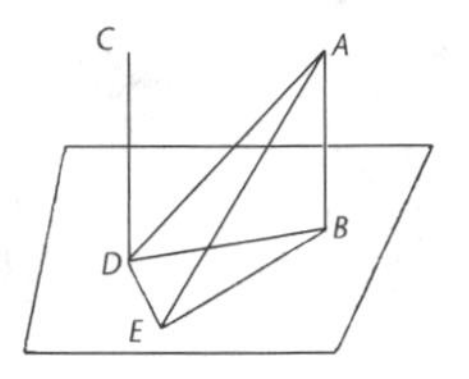

Proposition 7

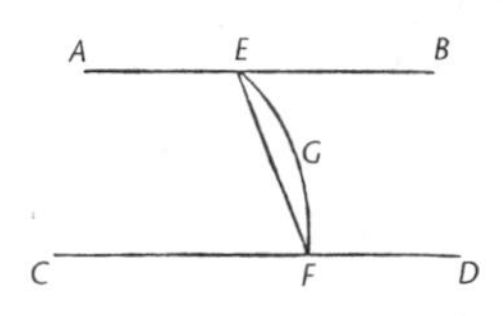

If two straight lines be parallel and points be taken at random on each of them, the straight line joining the points is in the same plane with the parallel straight line.

Proposition 8

If two straight lines be parallel, and one of them be at right angles to any plane, the remaining one will also be at right angles to the same plane.

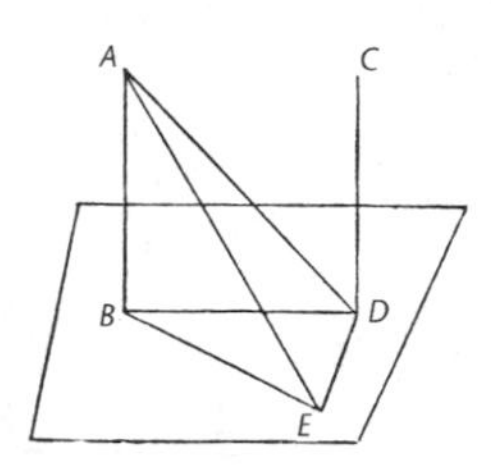

Proposition 9

Straight lines which are parallel to the same straight line and are not in the same plane with it are also parallel to one another.

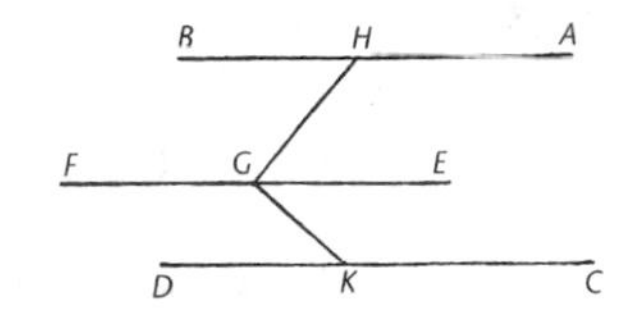

Proposition 10

If two straight lines meeting one another be parallel to two straight lines meeting one another not in the same plane, they will contain equal angles.

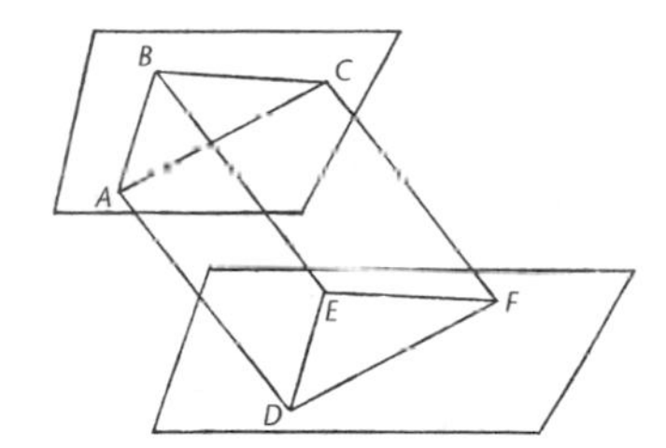

Proposition 11

From a given elevated point to draw a straight line perpendicular to a given plane.

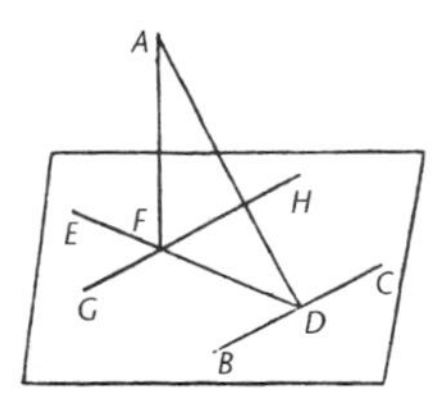

Proposition 12

To set up a straight line at right angles to a given plane from a given point in it.

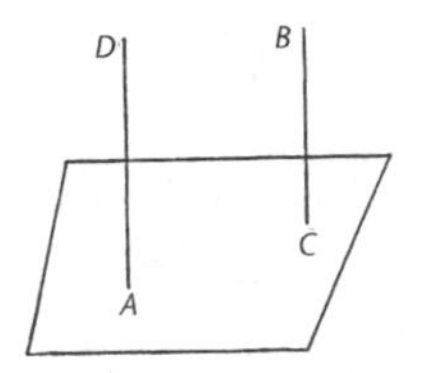

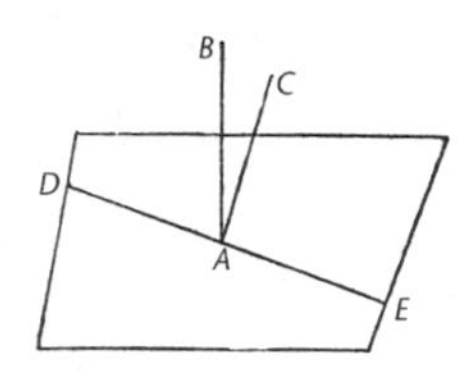

Proposition 13

From the same point two straight lines cannot be set up at right angles to the same plane on the same side.

Proposition 14

Planes to which the same straight line is at right angles will be parallel.

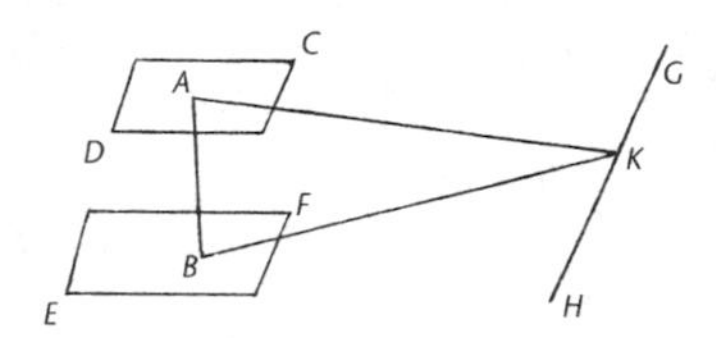

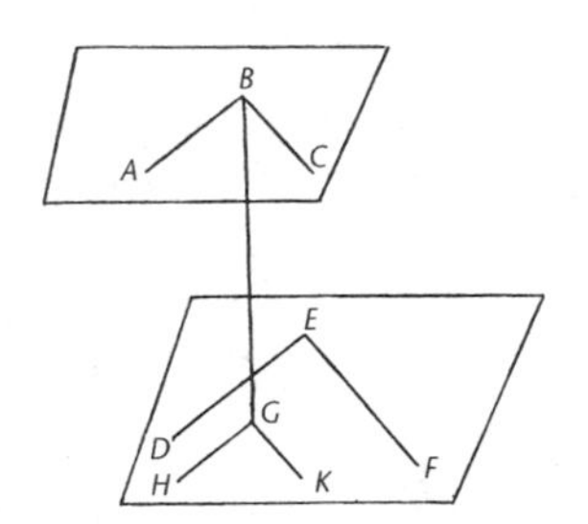

Proposition 15

If two straight lines meeting one another be parallel to two straight lines meeting one another, not being in the same plane, the planes through them are parallel.

Proposition 16

If two parallel planes be cut by any plane, their common sections are parallel.

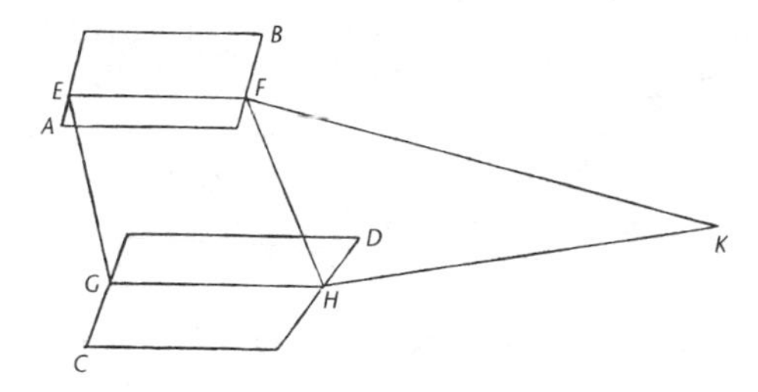

Proposition 17

If two straight lines be cut by parallel planes, they will be cut in the same ratios.

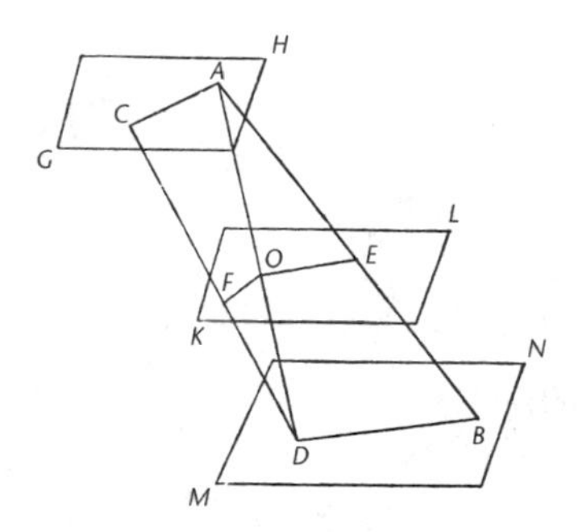

Proposition 18

If a straight line be at right angles to any plane, all the planes through it will also be at right angles to the same plane.

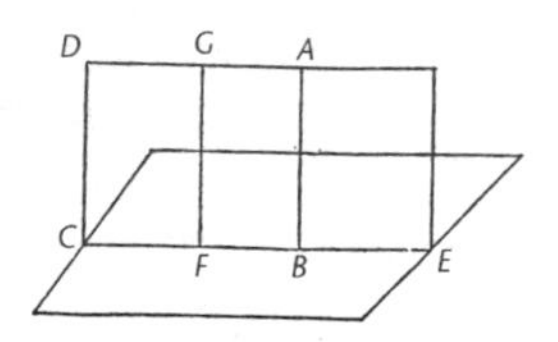

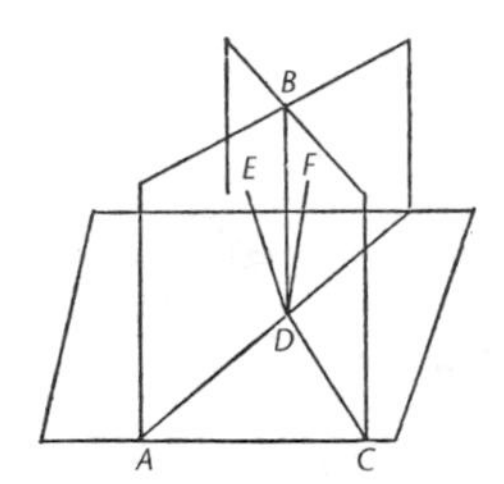

Proposition 19

If two planes which cut one another be at right angles to any plane, their common section will also be at right angles to the same plane.

Proposition 20

If a solid angle be contained by three plane angles, any two, taken together in any manner, are greater than the remaining one.

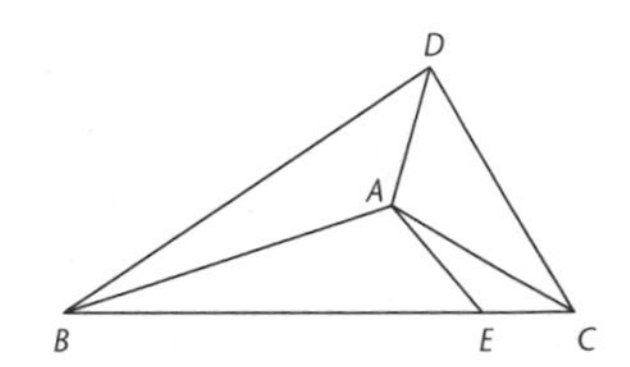

Proposition 21

Any solid angle is contained by plane angles less than four right angles.

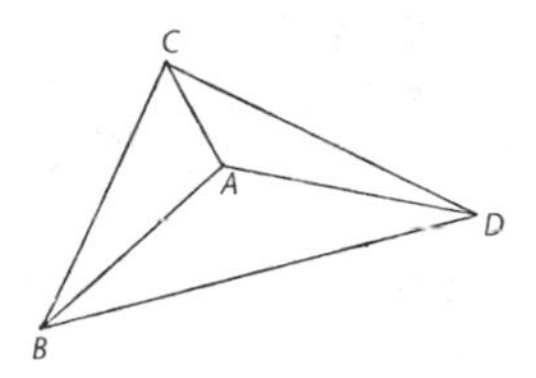

Proposition 22

If there be three plane angles of which two, taken together in any manner, are greater than the remaining one, and they are contained by equal straight lines, it is possible to construct a triangle out of the straight lines joining the extremities of the equal straight lines.

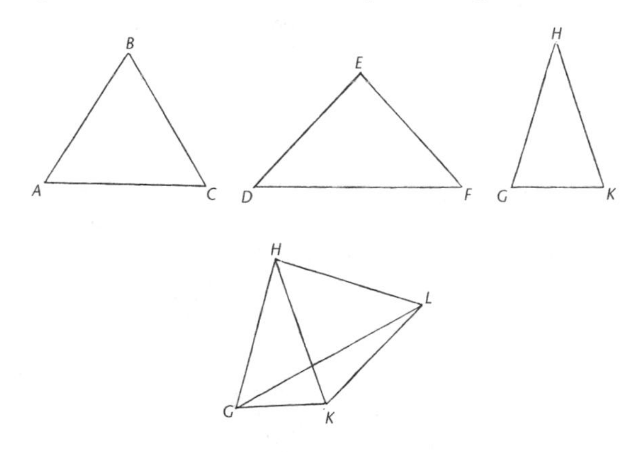

Proposition 23

To construct a solid angle out of three plane angles two of which, taken together in any manner, are greater than the remaining one: thus the three angles must be less than four right angles.

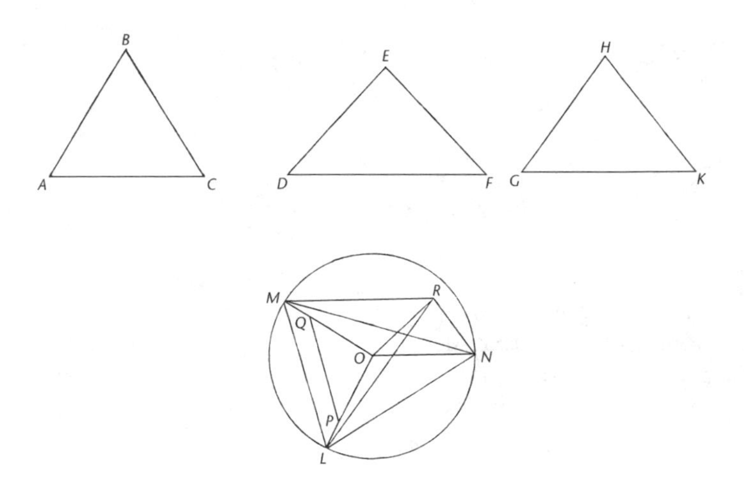

LEMMA. But how it is possible to take the square on *OR* equal to that area by which the square on *AB* is greater than the square on *LO,* we can show as follows.

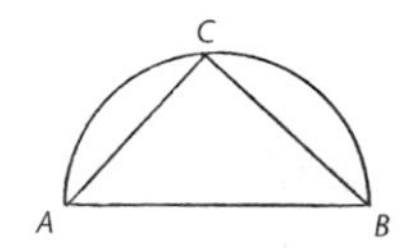

Proposition 24

If a solid be contained by parallel planes, the opposite planes in it are equal and parallelogrammic.

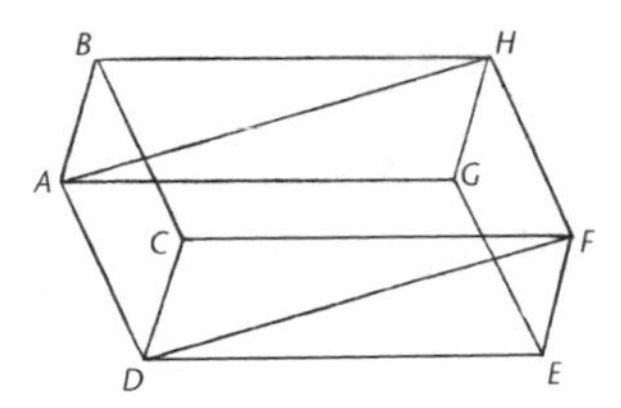

Proposition 25

If a parallelepipedal solid be cut by a plane which is parallel to the opposite planes, then, as the base is to the base, so will the solid be to the solid.

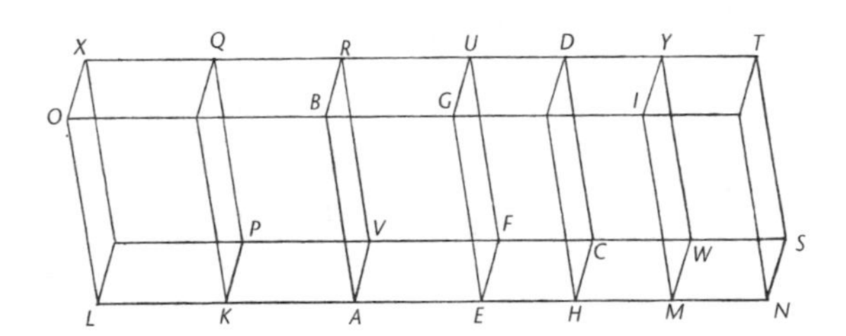

Proposition 26

On a given straight line, and at a given point on it, to construct a solid angle equal to a given solid angle.

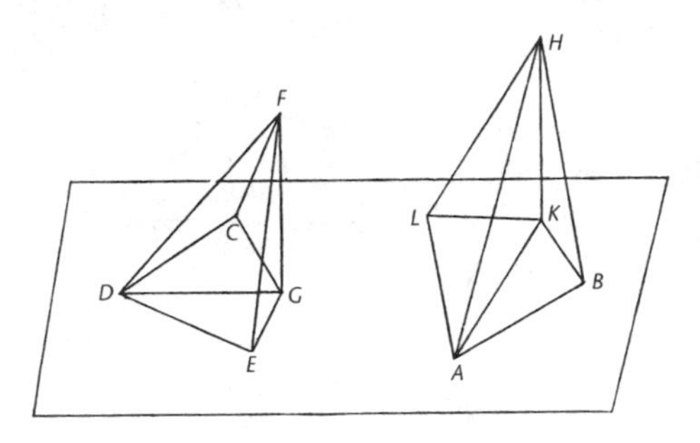

Proposition 27

On a given straight line to describe a parallelepipedal solid similar and similarly situated to a given parallelepipedal solid.

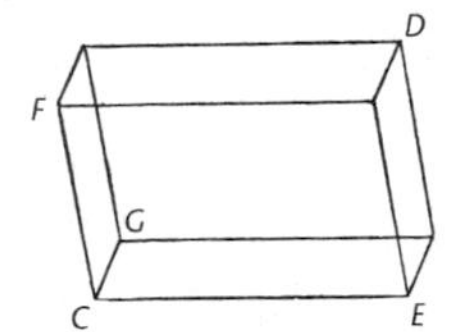

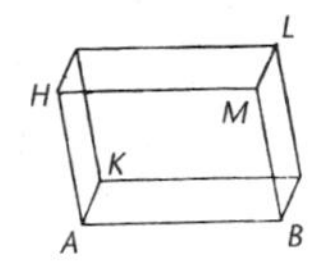

Proposition 28

If a parallelepipedal solid be cut by a plane through the diagonals of the opposite planes, the solid will be bisected by the plane.

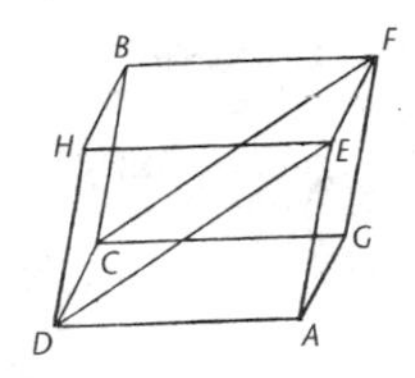

Proposition 29

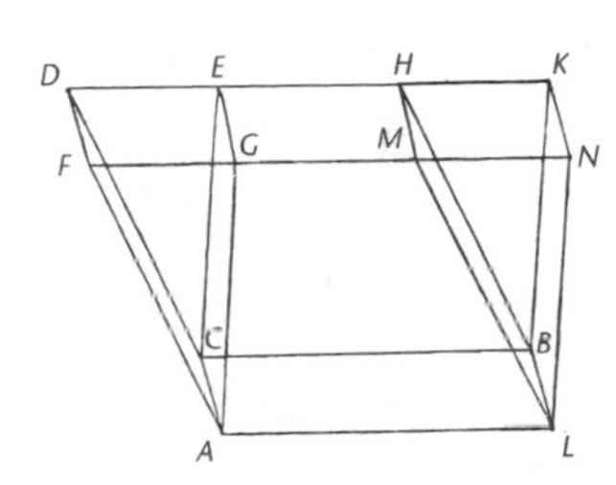

Parallelepipedal solids which are on the same base and of the same height, and in which the extremities of the sides which stand up are on the same straight lines, are equal to one another.

Proposition 30

Parallelepipedal solids which are on the same base and of the same height, and in which the extremities of the sides which stand up are not on the same straight lines, are equal to one another.

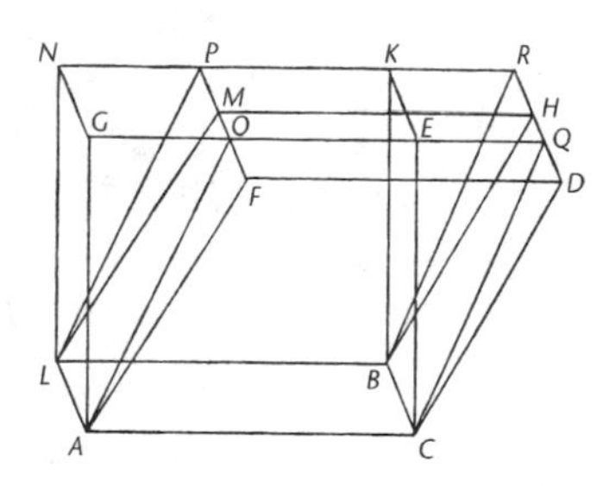

Proposition 31

Parallelepipedal solids which are on equal bases and of the same height are equal to one another.

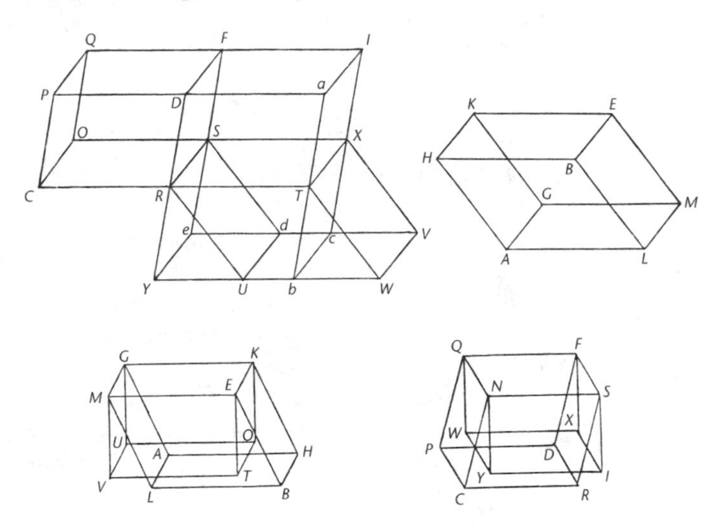

Proposition 32

Parallelepipedal solids which are of the same height are to one another as their bases.

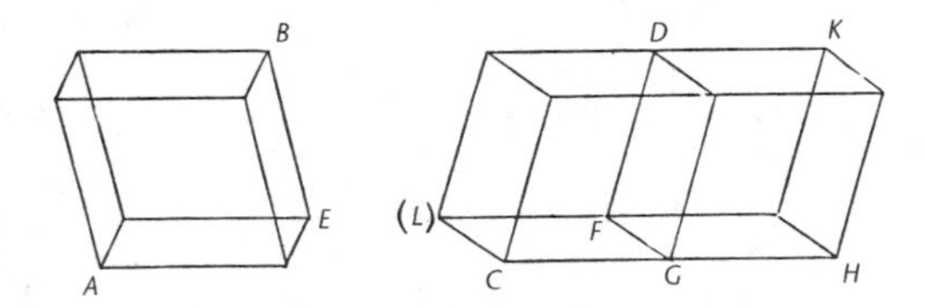

Proposition 33

Similar parallelepipedal solids are to one another in the triplicate ratio of their corresponding sides.

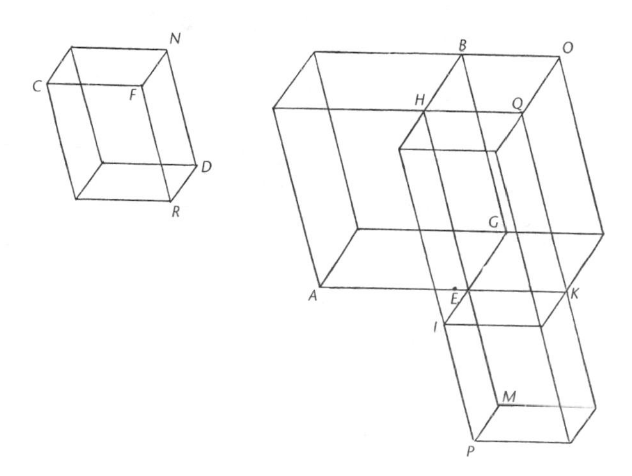

PORISM. From this it is manifest that, if four straight lines be [continuously] proportional, as the first is to the fourth, so will a parallelepipedal solid on the first be to the similar and similarly described parallelepipedal solid on the second, inasmuch as the first has to the fourth the ratio triplicate of that which it has to the second.

Proposition 34

In equal parallelepipedal solids the bases are reciprocally proportional to the heights; and those parallelepipedal solids in which the bases are reciprocally proportional to the heights are equal.

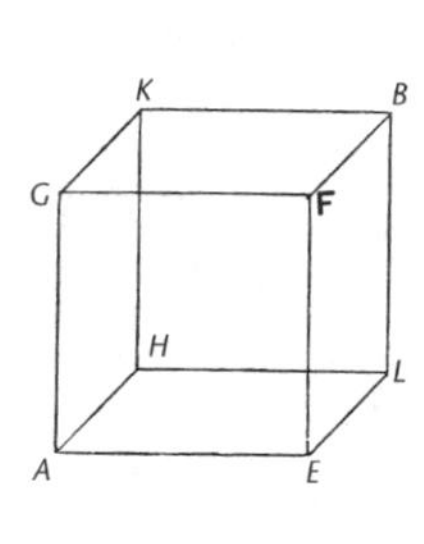

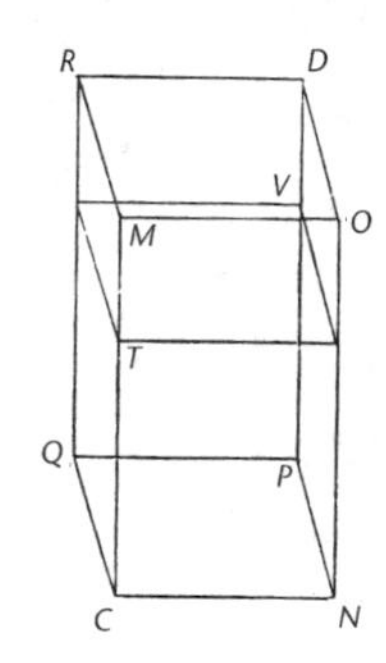

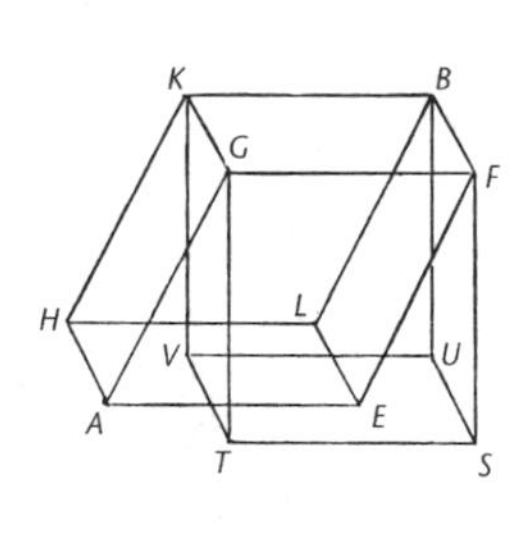

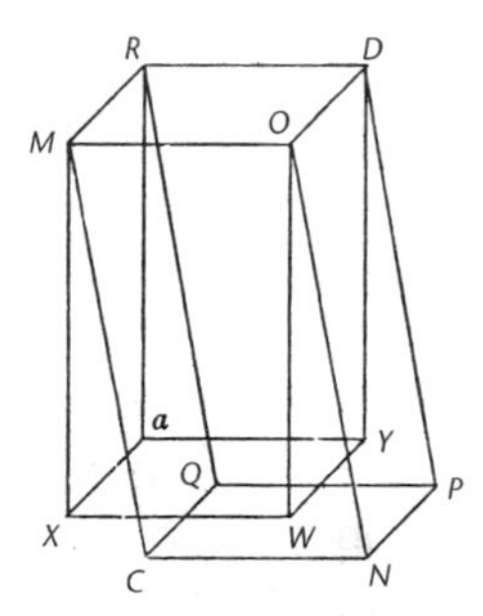

Proposition 35

If there be two equal plane angles, and on their vertices there be set up elevated straight lines containing equal angles with the original straight lines respectively, if on the elevated straight lines points be taken at random and perpendiculars be drawn from them to the planes in which the original angles are, and if from the points so arising in the planes straight lines be joined to the vertices of the original angles, they will contain, with the elevated straight lines, equal angles.

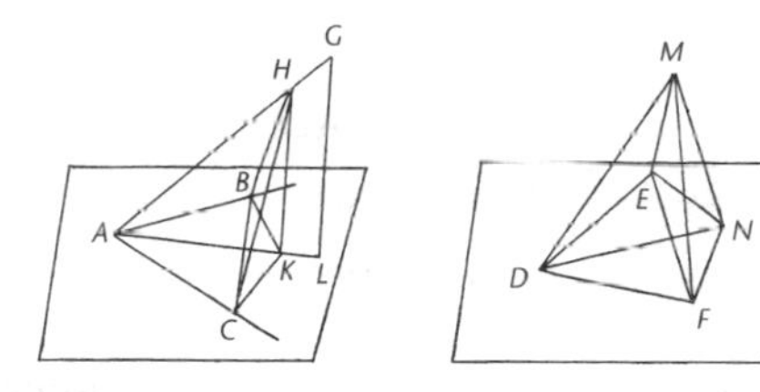

PORISM. From this it is manifest that, if there be two equal plane angles, and if there be set up on them elevated straight lines which are equal and contain equal angles with the original straight lines respectively, the perpendiculars drawn from their extremities to the planes in which are the original angles are equal to one another.

Proposition 36

If three straight lines be proportional, the parallelepipedal solid formed out of the three is equal to the parallelepipedal solid on the mean which is equilateral, but equiangular with the aforesaid solid.

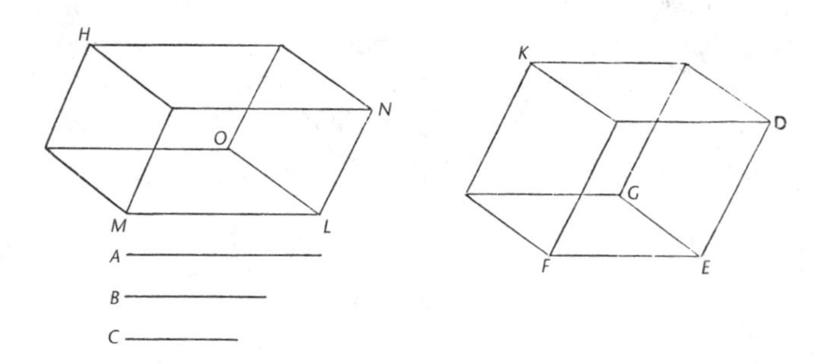

Proposition 37

If four straight lines be proportional, the parallelepipedal solids on them which are similar and similarly described will also be proportional; and, if the parallelepipedal solids on them which are similar and similarly described be proportional, the straight lines will themselves also be proportional.

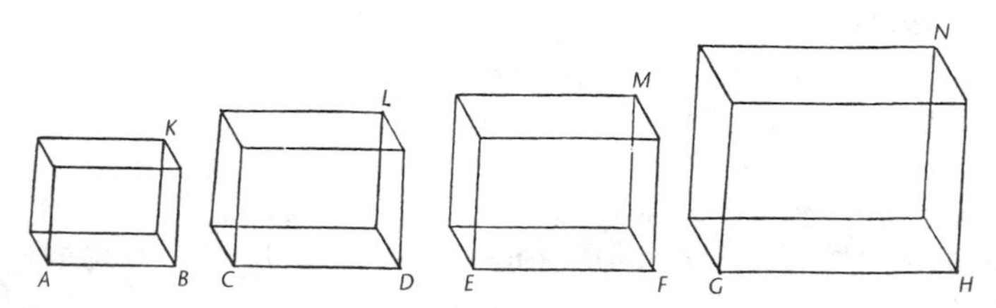

Proposition 38

If the sides of the opposite planes of a cube be bisected, and planes be carried through the points of section, the common section of the planes and the diameter of the cube bisect one another.

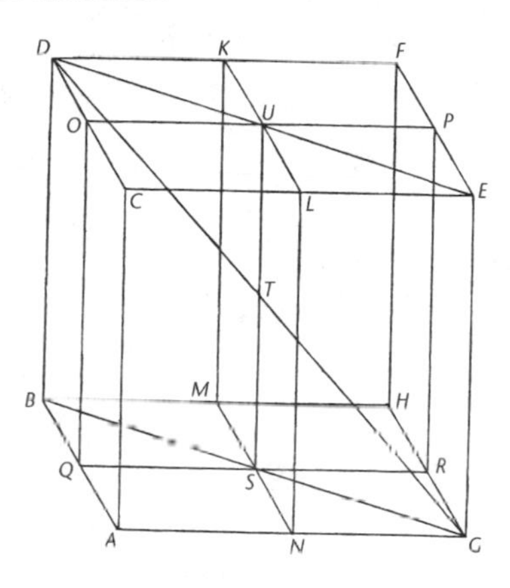

Proposition 39

If there be two prisms of equal height, and one have a parallelogram as base and the other a triangle, and if the parallelogram be double of the triangle, the prisms will be equal.

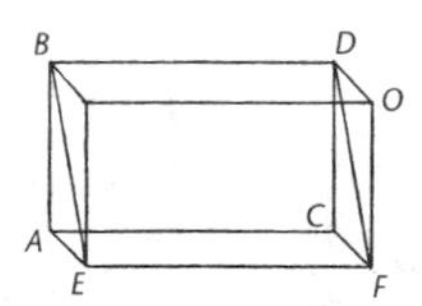

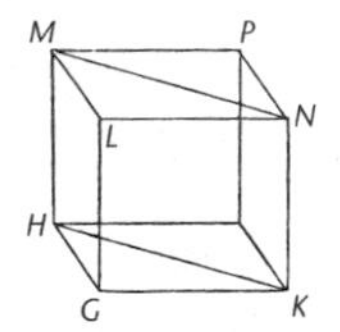

Book XII

Proposition 1

Similar polygons inscribed in circles are to one another as the squares on the diameters.

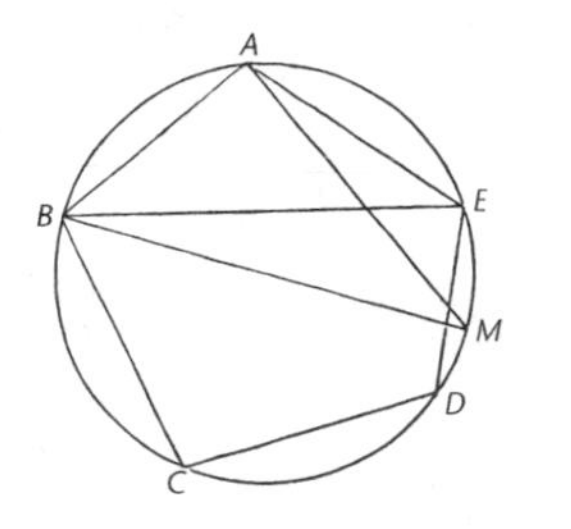

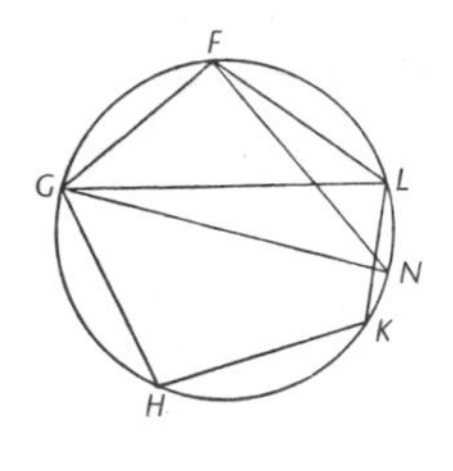

Proposition 2

Circles are to one another as the squares on the diameters.

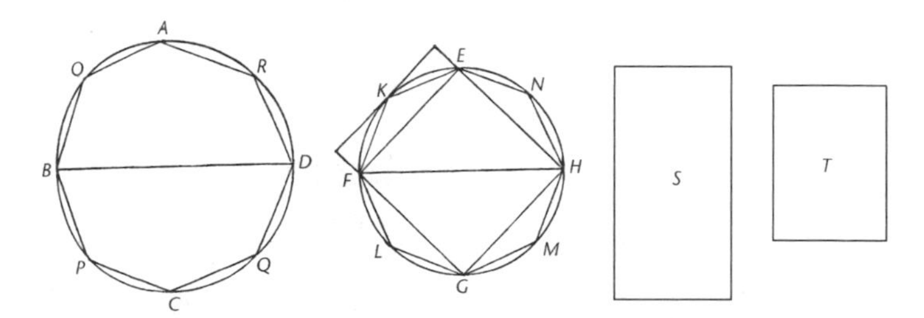

LEMMA. I say that, the area *S* being greater than the circle *EFGH,* as the area *S* is to the circle *ABCD,* so is the circle *EFGH* to some area less than the circle *ABCD.*

Proposition 3

Any pyramid which has a triangular base is divided into two pyramids equal and similar to one another, similar to the whole and having triangular bases, and into two equal prisms; and the two prisms are greater than the half of the whole pyramid.

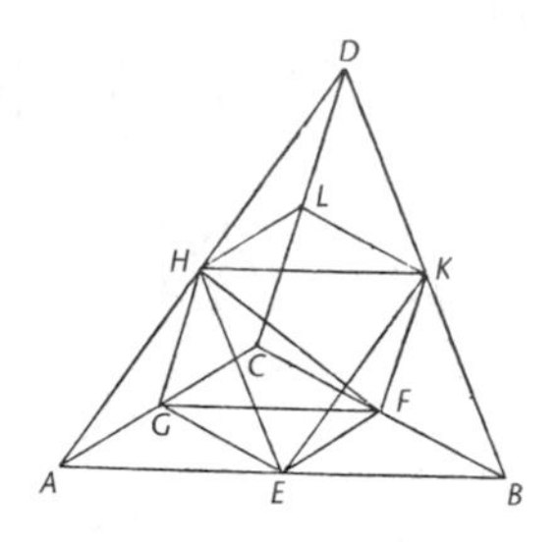

Proposition 4

If there be two pyramids of the same height which have triangular bases, and each of them be divided into two pyramids equal to one another and similar to the whole, and into two equal prisms, then, as the base of the one pyramid is to the base of the other pyramid, so will all the prisms in the one pyramid be to all the prisms, being equal in multitude, in the other pyramid.

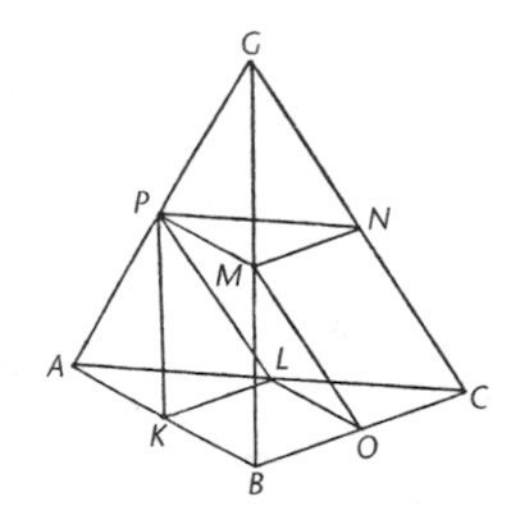

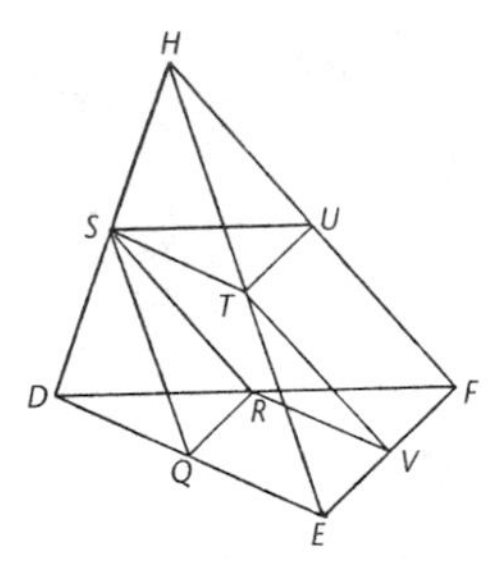

LEMMA. But that, as the triangle *LOC* is to the triangle *RVF*, so is the prism in which the triangle *LOC* is the base and

PMN its opposite, to the prism in which the triangle *RVF* is the base and *STU* its opposite, we must prove as follows.

Proposition 5

Pyramids which are of the same height and have triangular bases are to one another as the bases.

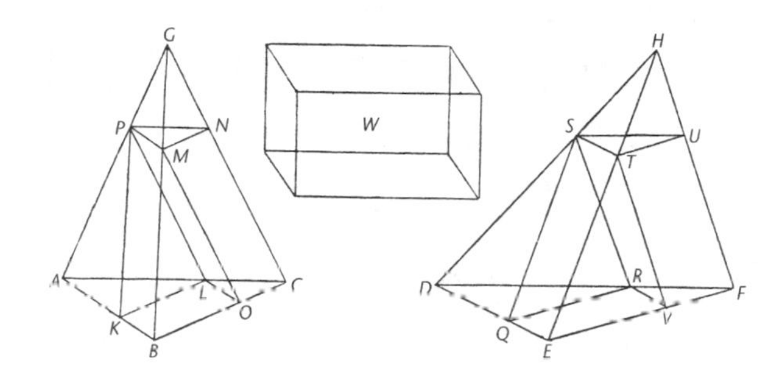

Proposition 6

Pyramids which are of the same height and have polygonal bases are to one another as the bases.

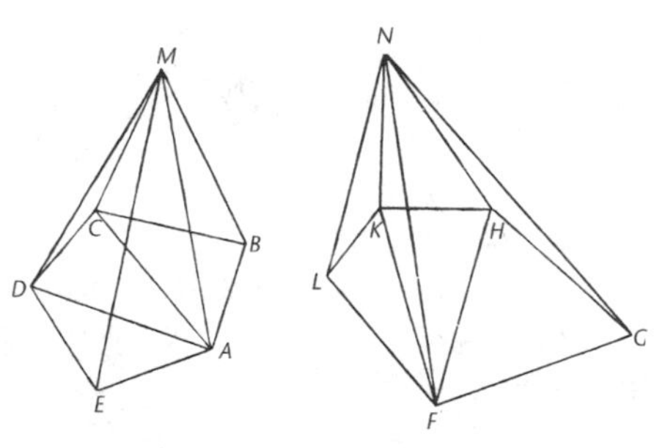

Proposition 7

Any prism which has a triangular base is divided into three pyramids equal to one another which have triangular bases.

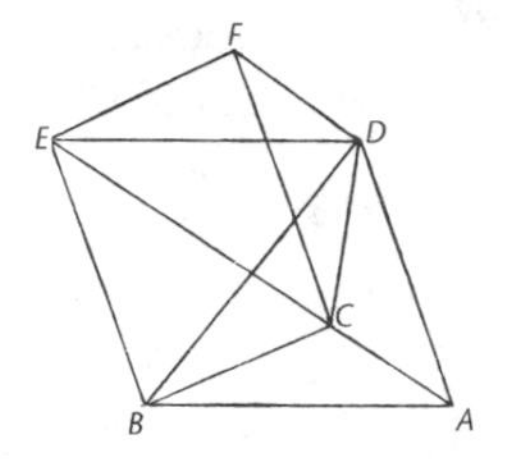

PORISM. From this it is manifest that any pyramid is a third part of the prism which has the same base with it and equal height.

Proposition 8

Similar pyramids which have triangular bases are in the triplicate ratio of their corresponding sides.

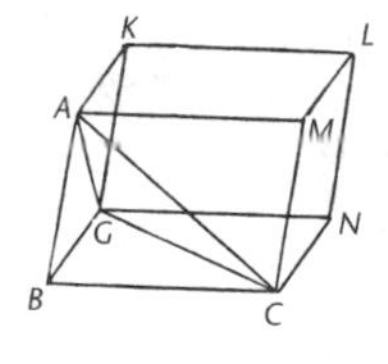

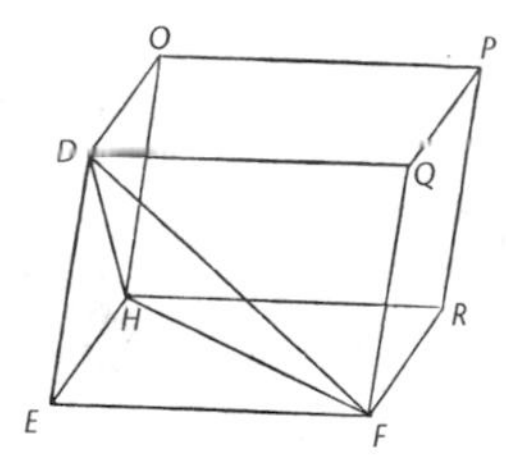

PORISM. From this it is manifest that similar pyramids which have polygonal bases are also to one another in the triplicate ratio of their corresponding sides.

Proposition 9

In equal pyramids which have triangular bases the bases are reciprocally proportional to the heights; and those pyramids in which the bases are reciprocally proportional to the heights are equal.

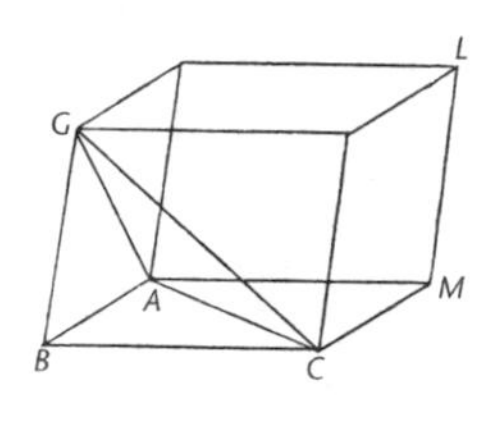

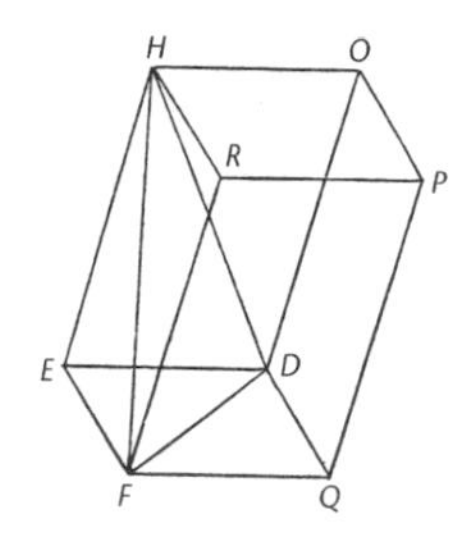

Proposition 10

Any cone is a third part of the cylinder which has the same base with it and equal height.

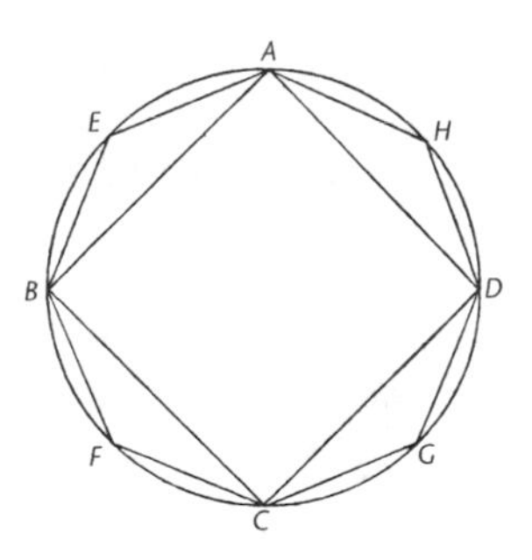

Proposition 11

Cones and cylinders which are of the same height are to one another as their bases.

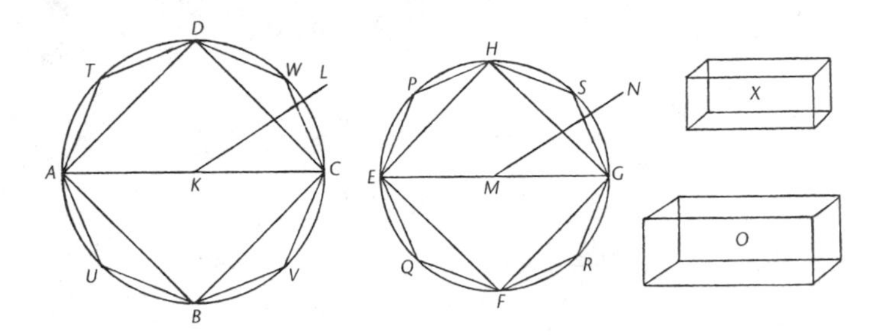

Proposition 12

Similar cones and cylinders are to one another in the triplicate ratio of the diameters in their bases.

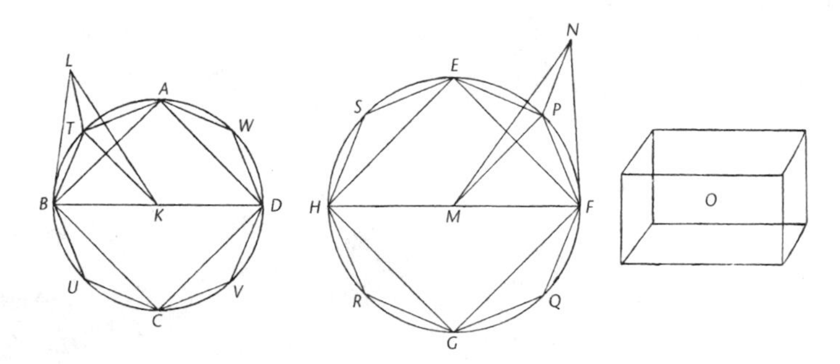

Proposition 13

If a cylinder be cut by a plane which is parallel to its opposite planes, then, as the cylinder is to the cylinder, so will the axis be to the axis.

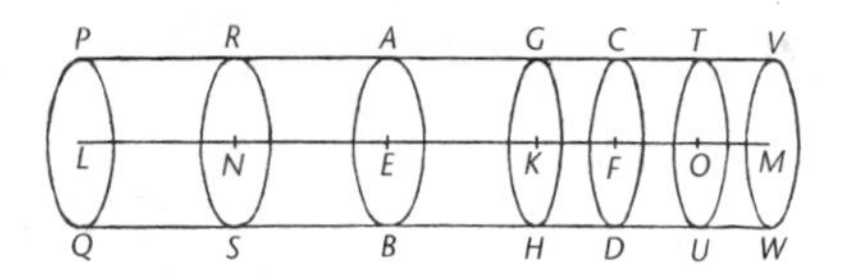

Proposition 14

Cones and cylinders which are on equal bases are to one another as their heights.

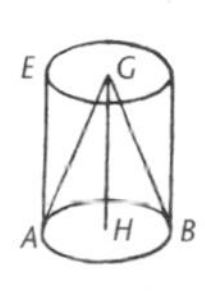

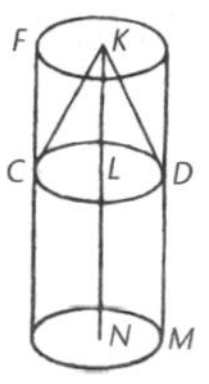

Proposition 15

In equal cones and cylinders the bases are reciprocally proportional to the heights; and those cones and cylinders in which the bases are reciprocally proportional to the heights are equal.

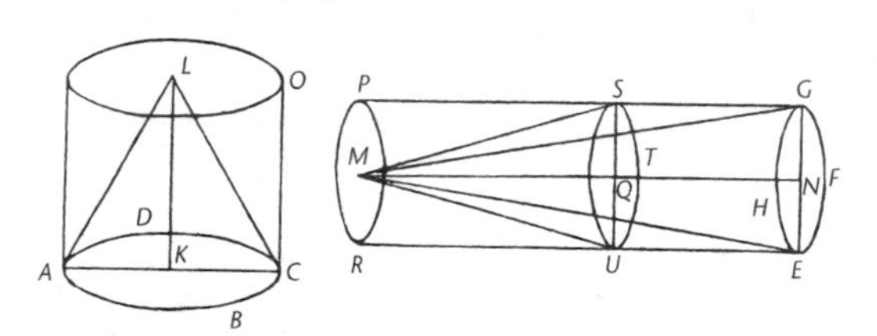

Proposition 16

Given two circles about the same centre, to inscribe in the greater circle an equilateral polygon with an even number of sides which does not touch the lesser circle.

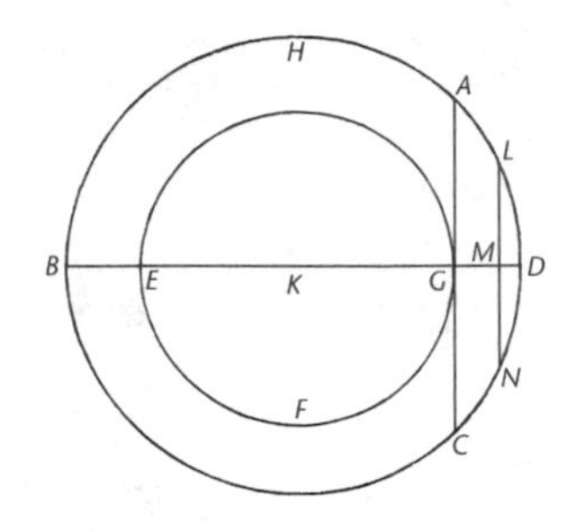

Proposition 17

Given two spheres about the same centre, to inscribe in the greater sphere a polyhedral solid which does not touch the lesser sphere at its surface.

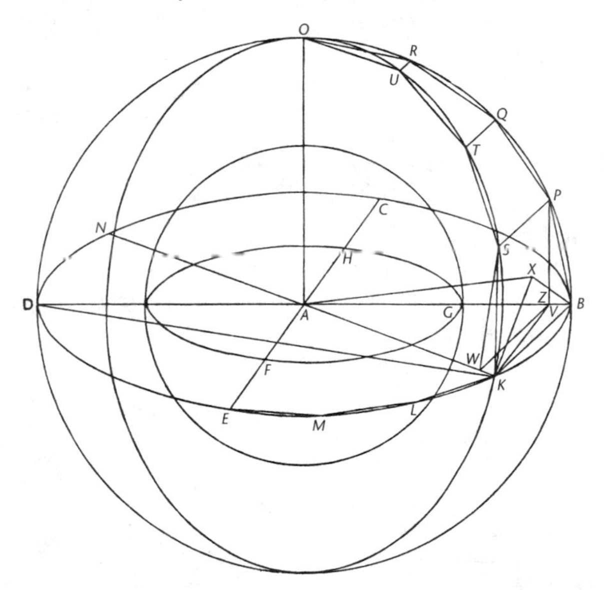

PORISM. But if in another sphere also a polyhedral solid be inscribed similar to the solid in the sphere *BCDE,* the polyhedral solid in the sphere *BCDE* has to the polyhedral solid in the other sphere the ratio triplicate of that which the diameter of the sphere *BCDE* has to the diameter of the other sphere.

Proposition 18

Spheres are to one another in the triplicate ratio of their respective diameters.

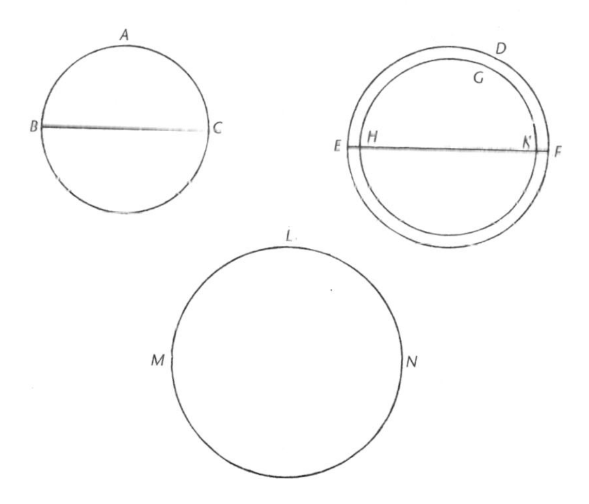

Book XIII

Proposition 1

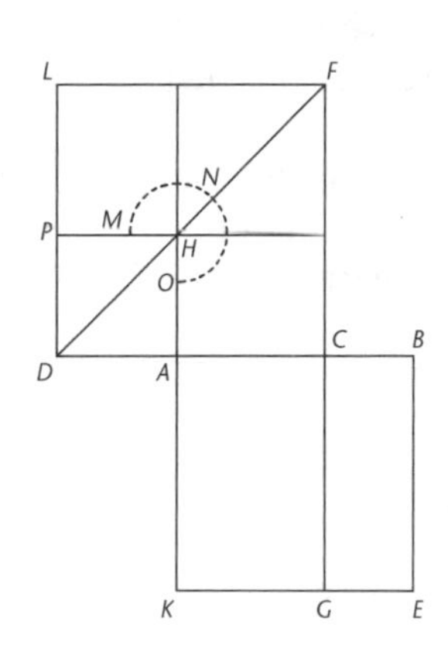

If a straight line be cut in extreme and mean ratio, the square on the greater segment added to the half of the whole is five times the square on the half.

Proposition 2

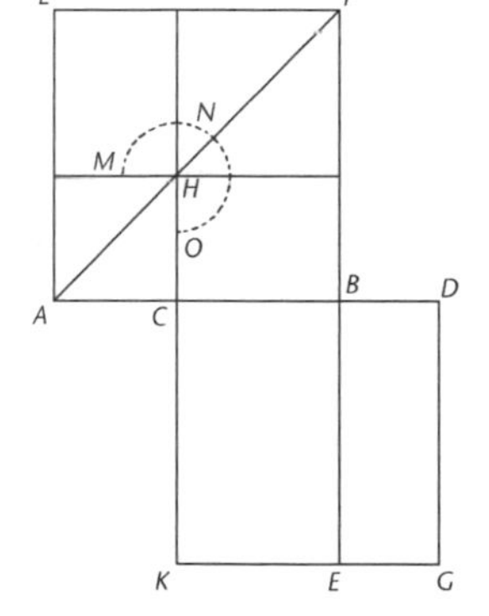

If the square on a straight line be five times the square on a segment of it, then, when the double of the said segment is cut in extreme and mean ratio, the greater segment is the remaining part of the original straight line.

LEMMA. That the double of *AC* is greater than *BC* is to be proved thus.

Proposition 3

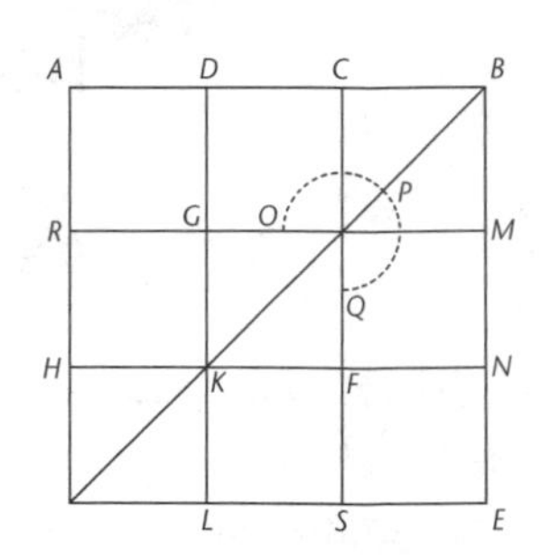

If a straight line be cut in extreme and mean ratio, the square on the lesser segment added to the half of the greater segment is five times the square on the half of the greater segment.

Proposition 4

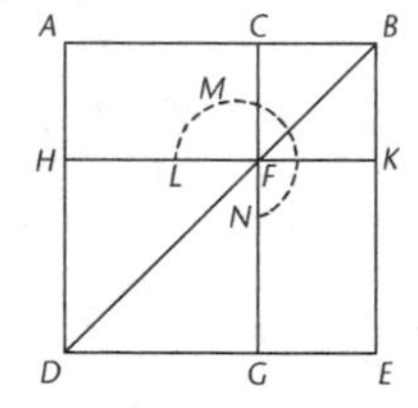

If a straight line be cut in extreme and mean ratio, the square on the whole and the square on the lesser segment together are triple of the square on the greater segment.

Proposition 5

If a straight line be cut in extreme and mean ratio, and there be added to it a straight line equal to the greater segment, the whole straight line has been cut in extreme and mean ratio, and the original straight line is the greater segment.

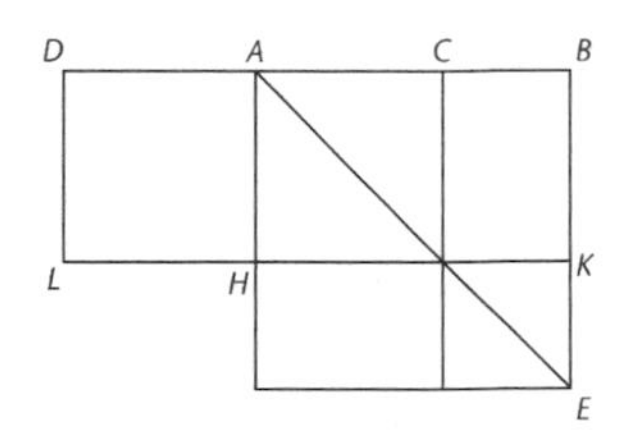

⟨Proposition 6

If a rational straight line be cut in extreme and mean ratio, each of the segments is the irrational straight line called apotome.⟩

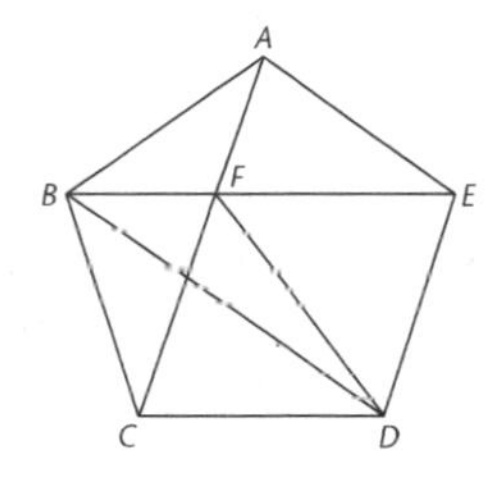

Proposition 7

If three angles of an equilateral pentagon, taken either in order or not in order, be equal, the pentagon will be equiangular.

Proposition 8

If in an equilateral and equiangular pentagon straight lines subtend two angles taken in order, they cut one another in extreme and mean ratio, and their greater segments are equal to the side of the pentagon.

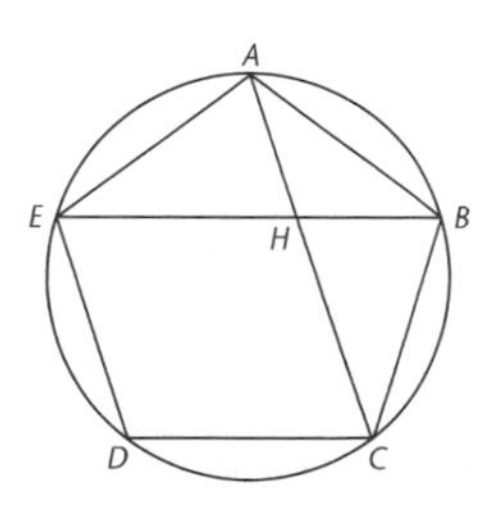

Proposition 9

If the side of the hexagon and that of the decagon inscribed in the same circle be added together, the whole straight line has been cut in extreme and mean ratio, and its greater segment is the side of the hexagon.

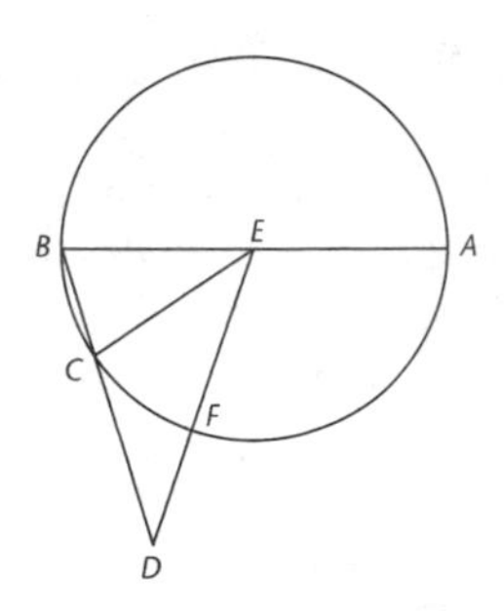

Proposition 10

If an equilateral pentagon be inscribed in a circle, the square on the side of the pentagon is equal to the squares on the side of the hexagon and on that of the decagon inscribed in the same circle.

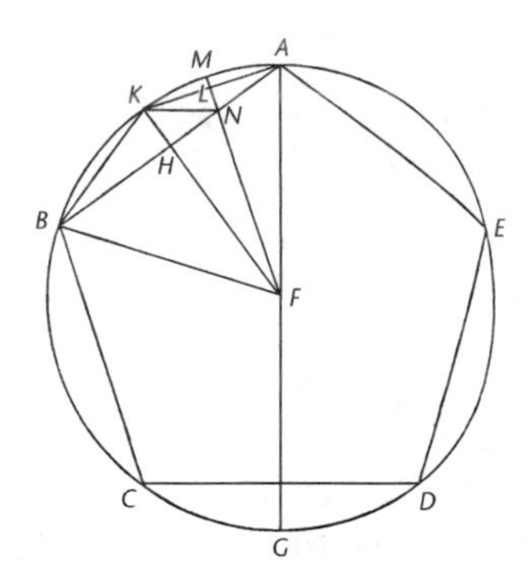

Proposition 11

If in a circle which has its diameter rational an equilateral pentagon be inscribed, the side of the pentagon is the irrational straight line called minor.

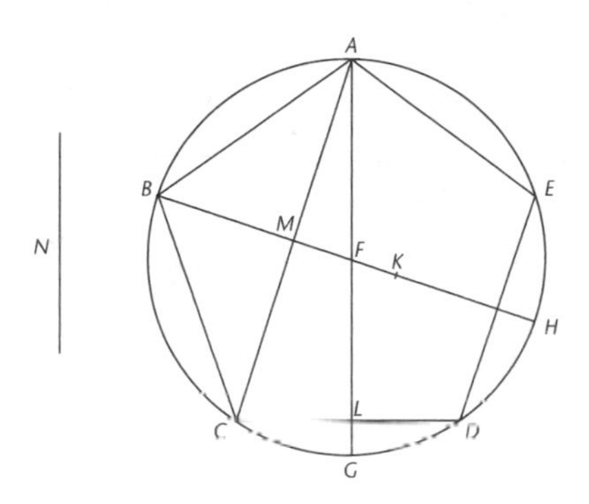

Proposition 12

If an equilateral triangle be inscribed in a circle, the square on the side of the triangle is triple of the square on the radius of the circle.

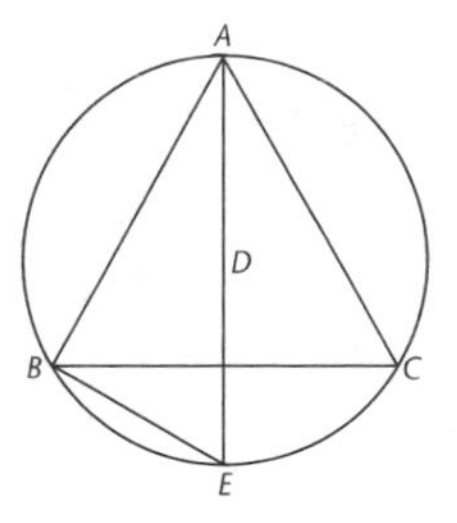

Proposition 13

To construct a pyramid, to comprehend it in a given sphere, and to prove that the square on the diameter of the sphere is one and a half times the square on the side of the pyramid.

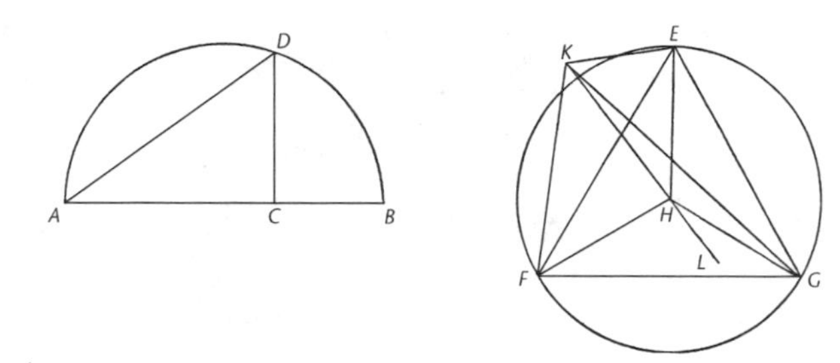

LEMMA. It is to be proved that, as *AB* is to *BC,* so is the square on *AD* to the square on *DC.*

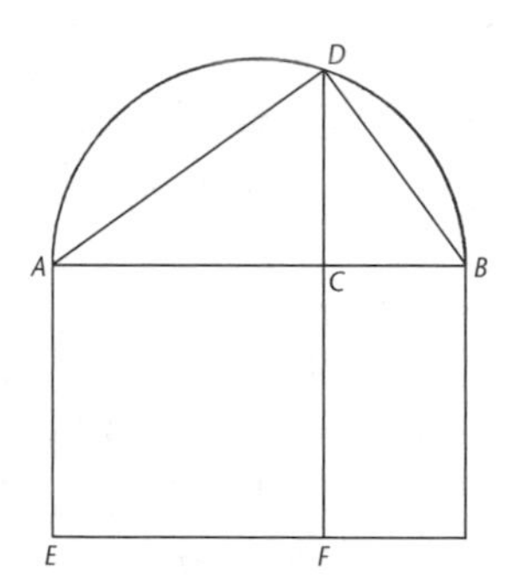

Proposition 14

To construct an octahedron and comprehend it in a sphere, as in the preceding case; and to prove that the square on the diameter of the sphere is double of the square on the side of the octahedron.

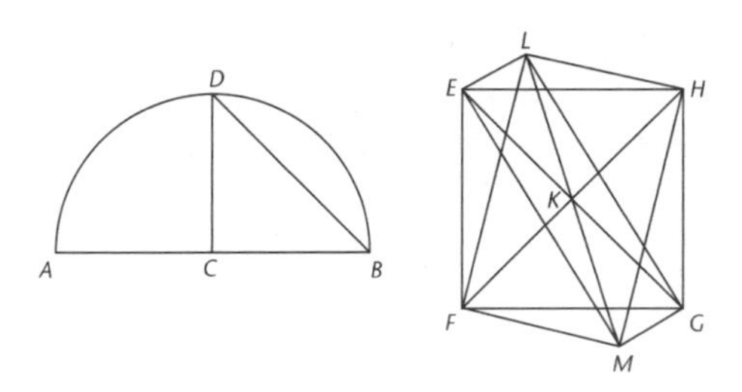

Proposition 15

To construct a cube and comprehend it in a sphere, like the pyramid; and to prove that the square on the diameter of the sphere is triple of the square on the side of the cube.

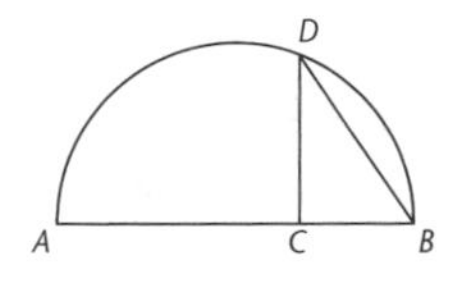

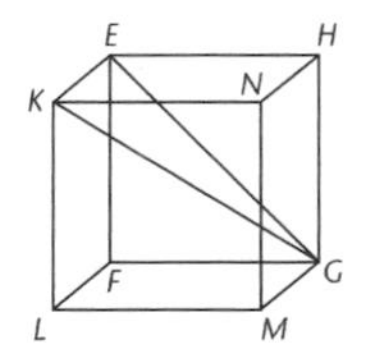

Proposition 16

To construct an icosahedron and comprehend it in a sphere, like the aforesaid figures; and to prove that the side of the icosahedron is the irrational straight line called minor.

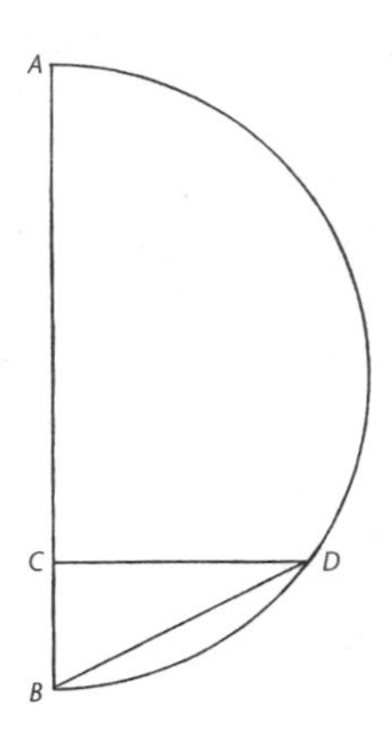

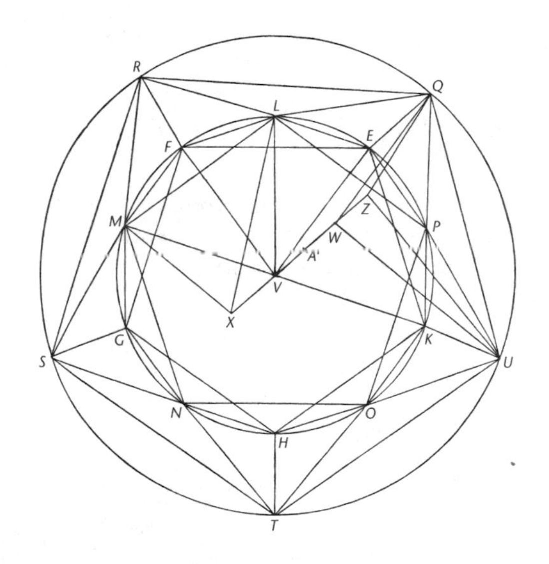

PORISM. From this it is manifest that the square on the diameter of the sphere is five times the square on the radius of the circle from which the icosahedron has been described, and that the diameter of the sphere is composed of the side of the hexagon and two of the sides of the decagon inscribed in the same circle.

Proposition 17

To construct a dodecahedron and comprehend it in a sphere, like the aforesaid figures, and to prove that the side of the dodecahedron is the irrational straight line called apotome.

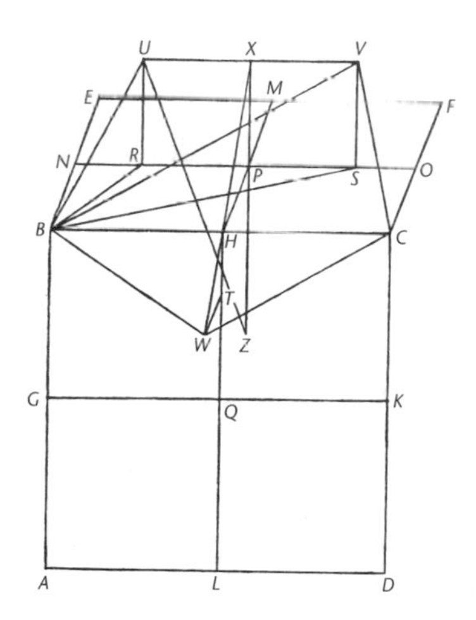

PORISM. From this it is manifest that, when the side of the cube is cut in extreme and mean ratio, the greater segment is the side of the dodecahedron.

Proposition 18

To set out the sides of the five figures and to compare them with one another.

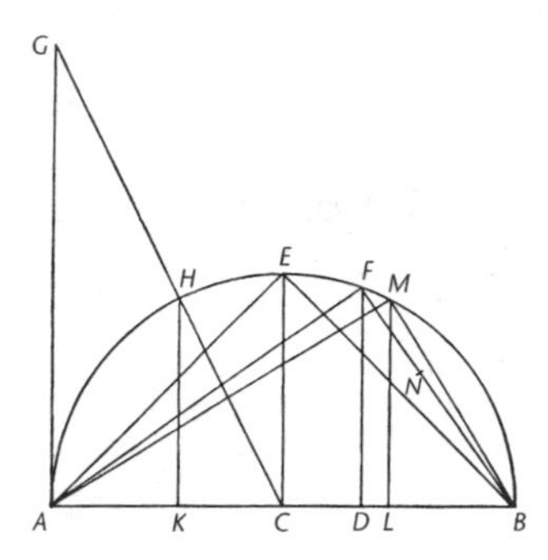

I say next that *no other figure, besides the said five figures, can be constructed which is contained by equilateral and equiangular figures equal to one another.*

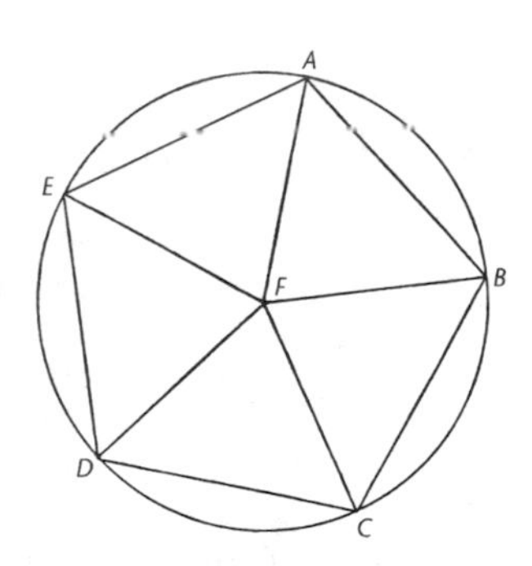

LEMMA. But that *the angle of the equilateral and equiangular pentagon is a right angle and a fifth* we must prove thus.